Human Resource Management in Construction

The construction sector is one of the most complex and problematic arenas within which to manage people. As a result, the applicability of much mainstream human resource management (HRM) theory to this industry is limited. Indeed, the operational realities faced by construction organizations mean that all too often the needs of employees are subjugated by performance concerns. This has potentially dire consequences for those who work in the industry, for the firms that employ them and ultimately, for the prosperity and productivity of the industry as a whole.

In this new edition of their leading text, Dainty and Loosemore have assembled a collection of perspectives which critically examine key aspects of the HRM function in the context of contemporary construction organizations. Rather than simply update the previous edition, the aim of this second edition is to provide a more critical commentary on the ways in which the industry addresses the HRM function and how this affects those who work within the industry. To this end, the editors have gathered contributions from many of the leading thinkers within construction HRM to critique the perspectives presented in the first edition. Each contributor either tackles specific aspects of the HRM function, or provides a critical commentary on industry practice. The authors explain, using real-life case studies, the ways in which construction firms respond to the myriad pressures that they face through their HRM practices.

Together the contributions encourage the reader to rethink the HRM function and its role in defining the employment relationship. This provides essential reading for students of construction and project management, and reflective practitioners who are interested in theoretically informed insights into industry practice and its implications.

Andrew Dainty is Professor of Construction Sociology at Loughborough University, UK.

Martin Loosemore is Professor of Construction Management and Associate Dean at the University of New South Wales, Australia.

Human Resource Management in Construction

Critical perspectives

Second edition

Edited by Andrew Dainty and Martin Loosemore

Routledge
Taylor & Francis Group

LONDON AND NEW YORK

First published 2012
by Routledge
2 Park Square, Milton Park, Abingdon, Oxon OX14 4RN

Simultaneously published in the USA and Canada
by Routledge
711 Third Avenue, New York, NY 10017

Routledge is an imprint of the Taylor & Francis Group, an informa business

British Library Cataloguing in Publication Data
A catalogue record for this book is available from the British Library

Library of Congress Cataloging in Publication Data
Human resource management in construction : critical perspectives / edited by Andrew Dainty and Martin Loosemore.
p. cm.

Rev. ed. of: Human resource management in construction projects : strategic and operational approaches / Martin Loosemore, Andrew Dainty, and Helen Lingard. c2003.
Includes bibliographical references and index.

1. Construction industry--Personnel management. I. Dainty, Andrew. II. Loosemore, Martin, 1962- Human resource management in construction projects.

HD9715.A2L656 2012

624.068'3--dc23

2011045601

ISBN: 978-0-415-59306-9 (hbk)
ISBN: 978-0-415-59307-6 (pbk)
ISBN: 978-0-203-84247-8 (ebk)

Typeset in Sabon
by Saxon Graphics Ltd, Derby

Printed and bound by CPI Group (UK) Ltd, Croydon, CR0 4YY

Contents

Tables

Figures

Case Studies

Dedication

We dedicate this book to the memory of our dear friend and colleague Professor David Langford (1950–2010). Dave was in the process of preparing a chapter for this book when he sadly passed away in 2010. An inspiration and mentor to many and forever an advocate of the 'counter-vailing view', Dave's enthusiasm for a text which questioned the norms of practice within our industry was a key driver in developing this book. We hope that the range of critical perspectives and positions gathered here in some way reflect the values for which he was known and respected by so many within the construction management research community.

Preface

When we set out to develop the first edition of this book we had a clear orientation on developing a text which provided theoretically informed, but practical advice to construction practitioners on how they could manage their human resources more strategically. The book had an overtly performative orientation in that it sought to identify practices which could help deliver strategic goals. As we stated in our preface to the first edition '*If we can persuade project managers of the strong link that exists between HRM performance and project performance, we will have succeeded in our aim*'.

At that time we perceived there to be a need to combine our understanding of construction organisations and projects with established perspectives on managing people from the mainstream HRM literature. At that time, HRM research within the construction research community was fairly nascent, with most studies firmly anchored in supporting the dominant discourse of the need for radical industry improvement. Leading texts from the mainstream HRM field were similarly replete with strategies for maximising the value of people to organisations, and of aligning the behaviours of organisational members with corporate objectives. We sought to emulate this approach through the application of such thinking to the construction industry context. Arguably, such a perspective was valuable at the time, especially as it enabled us to foreground the importance of people management within the performance improvement agenda.

Almost a decade on from writing the first edition, we would argue that a new set of challenges now confront the people management function in the industry, challenges that require a different set of perspectives and reflections. For example, industry reforms have been implemented in many countries; a global financial crisis has affected the priorities and strategies of many construction organisations; productivity improvement has become a priority in many countries; partnership approaches, alliance contracting and integrated supply chains have become more widespread; some countries have suffered from massive skills shortages (and a subsequent reliance on migrant workers), whilst others are facing considerable skills oversupply.

At the same time, in academia a more critical perspective on people management practice has emerged which sits alongside those which emphasise the need for continual improvement in performance. These more critical perspectives on HRM have begun to influence and shape the construction research literature, and arguably provide a set of interesting counterpoints to the perspectives which dominated in the 1990s.

In this edition we have sought to both recognise the importance of HRM in supporting industry development and performance, and also the need to *challenge* industry practice and to induce new lines of thinking about people management in the industry. In order to do this we have enlisted the contributions of leading researchers from built environment and business schools around the world who are renowned for their critical insights into how to better understand and challenge current construction industry practices. We have deliberately coalesced contributions from scholars who have, through their own research and practice insights, questioned some of the established orthodoxies within the sector. Others are leading researchers from the HR field whose work has shaped thinking across many industries and sectors. Rather than being prescriptive, we have encouraged our authors to take their chapter in whichever direction they feel contributes to debates within their field – directly challenging issues raised in the first edition of this book where appropriate. In approaching this edition in this way, we hope that this book builds on the first edition by critiquing and extending current debates within our own research community and the construction industry at large. The new title also reflects this orientation.

All of the chapters are based on contemporary research and have been written in a clear, easy-to-read style. Many use case studies and vignettes to illustrate their perspectives and/or to ground their commentaries on existing industry practice. Together, the contributions reveal both the intended and unintended consequences of HRM to both those who work in the industry and the organisations which employ them. The chapters provide a fresh set of reflections, provocations and potential trajectories for both the industry and the research community to explore in future. We hope that you enjoy them as much as we have.

Andy Dainty and Martin Loosemore
Loughborough, UK and Sydney, Australia

About the Contributors

Andrew Dainty is Professor of Construction Sociology and Associate Dean (Research) within the School of Civil and Building Engineering, Loughborough University. His research focuses on human social action within construction and other project-based sectors and particularly the social rules and processes that affect people working as members of project teams. He also leads work mobilising critical perspectives on project failure and performance. He has published over 250 papers in academic journals and conferences and is co-author/editor of seven books and research monographs. Andrew is co-editor of *Construction Management and Economics*, a past chair of the *Association of Researchers in Construction Management* (ARCOM) and joint coordinator of CIB TG 76 on Recognising Innovation in construction. He is currently Visiting Professor at Chalmers University (Gothenburg, Sweden) and is Conjoint Professor at The University of Newcastle (NSW, Australia).

Martin Loosemore is Professor of Construction Management at the University of New South Wales, Sydney, Australia. He is a Fellow of the Royal Institution of Chartered Surveyors and a Fellow of the Chartered Institute of Building and a visiting Professor at University of Loughborough, UK. Martin was an advisor to the Australian Royal Commission into the Building Industry, the Federal Senate inquiry into the Building and Construction Industry and is a member the Australian Government's Built Environment Innovation Council which advises the Federal Australian government on innovation policy and reform.

Don Aroney is Brookfield Multiplex's Executive Director Operations. He oversees Brookfield Multiplex's operational excellence objectives and has ultimate responsibility for all aspects of Brookfield Multiplex operations, promoting excellence across the entire operational system including delivery, finance, marketing, customer satisfaction and people management. Prior to his current role, Don was a Regional Director and Site Manager on the construction of the Sydney Olympic Stadium and was responsible for the implementation of sweeping safety initiatives across the business.

Paul Chan is Lecturer in Project Management at the School of Mechanical, Aerospace and Civil Engineering, University of Manchester. He actively studies Human Resource Management (HRM) and Organisational Development issues in project-based contexts. He has worked with the construction, manufacturing and aviation sectors. He co-authored *Constructing Futures: Industry Leaders and Futures Thinking in Construction*, and has written over 50 peer-reviewed academic journal and conference articles. He is Secretary of the Association of Researchers in Construction Management (ARCOM) in the UK, and joint coordinator of CIB Task Group 78 on 'Informality and Emergence in Construction'.

Linda Clarke is Professor of European Industrial Relations in the Westminster Business School, University of Westminster and Director of the Centre for the Study of the Production of the Built Environment (ProBE). She has long experience of comparative research on labour, vocational education, skills, and wage relations in Europe and particular expertise in the construction sector, being also on the board of the European Institute for Construction Labour Research (CLR). Her most recent publication is: (2011) *Knowledge, Skills, Competence in the European Labour Market: What's in a Qualification?* with Brockmann, Winch *et al.*, Oxford, Routledge.

Janet Druker joined Westminster University in 2010 to work as a Visiting Professor with the team in ProBE (Research Centre for the Production of the Built Environment). She extended that link in 2011 and is based in the Human Resource Management Department of Westminster Business School. She is also an Emeritus Professor of Canterbury Christ Church University where she was Senior Pro-Vice Chancellor between 2005 and 2010. Her research interests are in the field of work and employment, with particular reference to resourcing and reward issues.

Stuart Green is Professor of Construction Management and Head of the School of Construction Management and Engineering at the University of Reading. He has extensive experience of research leadership having been responsible for managing in excess of £7 million of funded research. Stuart has published extensively in leading international journals and is especially well known for his critical perspectives on a range of topics relating to construction. He enjoys extensive policy connectivity within the UK construction sector, as evidenced by his roles as Core Commissioner on the Commission for a Sustainable London 2012. He is currently chairman of the Chartered Institute of Building's Innovation and Research Panel.

Dave Higgon is co-author of *Risk Management in Projects* with extensive experience in the Australian Building and Construction Industry. He is currently working as a consultant to industry and lectures part-time in OHS at the University of New South Wales. An influential proponent of 'cooperative culture between management and the workforce', Dave has used this premise to develop innovative and successful workplace solutions in the areas of Industrial Relations, Occupational Health and Safety and Training and Development. His background is varied spanning work as a tradesperson, subcontractor, Building Union Official, and Safety Manager. He has most recently performed the role of Employee Relations Manager for a major construction company.

Stewart Johnstone is lecturer in Human Resource Management at Newcastle University Business School, UK. Prior to this he worked at the School of Business and Economics at Loughborough University. His research interests traverse human resource management and employment relations, but current interests include employee voice and representation, HRM strategy, and studies of HRM in different sectoral and institutional contexts. He is a Chartered Member of the Chartered Institute of Personnel and Development (CIPD).

Helen Lingard is a Professor in the School of Property, Construction and Project Management, RMIT University, where she leads a team of researchers in the field of construction safety and health. Helen was awarded a PhD by the University of Hong Kong and worked as a safety and health advisor for Costain Building and Civil Engineering in the delivery of major infrastructure construction projects. In Australia, Helen has provided safety and health consultancy services to the construction, mining and telecommunications industries. Helen has undertaken numerous contract research projects for industry and government. Her health and safety research is currently funded by the Australian Research Council under the Future Fellowship Scheme (Grant no. FT0990337).

Charlie McGuire is a former member of the Union of Construction, Allied Trades and Technicians (UCATT), and is currently a research fellow in the Centre for the Study of the Production of the Built Environment (ProBE), University of Westminster. He has researched and published mainly in the field of Irish labour history. His most recent publication is *Sean McLoughlin: Ireland's Forgotten Revolutionary*, 2011, Pontypool, Merlin Press.

Mick Marchington is Emeritus Professor of Human Resource Management at Manchester Business School, University of Manchester. He has held visiting posts at the Universities of Sydney, Auckland and Paris. He has published widely on Human Resource Management (HRM), both in books and in refereed journals. He is best known for his work on employee involvement and participation and on HRM across organisational boundaries. He is Co-Editor of the Human Resource Management Journal (HRMJ) and is a Chartered Companion of the CIPD.

Kate Ness joined the School of Construction Management and Engineering at the University of Reading in 2007, having worked previously for more than 25 years in the construction industry. She was initially trained as a bricklayer and hence has first-hand experience of the workplace challenges faced by construction operatives. Kate's research interests relate to people and culture in construction, with a particular focus on skill and identity in the building trades. Her previous work includes the use of critical discourse analysis to identify some of the assumptions behind the text of the 'Respect for People' report on UK construction, particularly relating to safety and diversity. Kate has published widely on gender issues in the construction sectors, and is especially critical of the cliché that 'most jobs in construction can be done by women'.

Abigail Powell is a Research Associate at the Social Policy Research Centre (SPRC) at the University of New South Wales, Australia. Abigail's research interests are in social policy, equity and diversity. Prior to joining SPRC in 2009, Abigail worked at Loughborough University in the UK where she worked on research funded by the Economic and Social Research Council, the European Commission and the UK Resource Centre for Women in Science, Engineering and Technology. This research investigated women's experiences of working in male-dominated environments, particularly engineering and construction.

Ani Raidén is a senior lecturer in Human Resource Management at Nottingham Business School, Nottingham Trent University. She holds a PhD from Loughborough University and is a Chartered Member of the CIPD. Her research on employee resourcing, human resource development and work–life balance in the construction industry has been published in both construction and human resource management journals and conferences. Ani co-authored *Employee Resourcing in the Construction Industry (2009)* with Andrew Dainty and Richard Neale.

Katherine Sang is a research fellow at the Centre for Diversity and Equality in Careers and Employment Research at the University of East Anglia. Her research interests include gender in the construction industry, women's careers in academia and participatory research. She is on the Executive of the Feminist and Women's Studies Association and is co-chair of ResNet, a networking group for women researchers at the University of East Anglia.

Anne Sempik is a principal lecturer in Human Resource Management and programme leader for MSc Strategic HRM at Nottingham Business School, Nottingham Trent University. She holds a PhD from Nottingham University and is a Chartered Fellow of the CIPD. Her academic and professional interests focus on talent management, assessment and selection, succession planning and the impact of culture on teaching and learning. Prior to her academic career, Anne worked as a people resourcing specialist in a large private sector organisation.

Christine Wall is senior research fellow at the Centre for the Study of the Production of the Built Environment (ProBE), University of Westminster. She has spent many years researching and publishing on training, skill, the role of labour and particularly the role of women, in the history of the construction industry. Her most recent publication is *Work and Identity: historical and cultural contexts,* with John Kirk, 2010, Basingstoke: Palgrave Macmillan.

Adrian Wilkinson is Professor and Director of the Centre for Work, Organisation and Wellbeing at Griffith University, Australia. Prior to his 2006 appointment, Adrian worked at Loughborough University in the UK where he was Professor of Human Resource Management from 1998, and Director of Research for the Business School. Adrian has also worked at the Manchester School of Management at the University of Manchester Institute of Science and Technology. He holds Visiting Professorships at Loughborough University, Sheffield University and the University of Durham, and is an Academic Fellow at the Centre for International Human Resource Management at the Judge Institute, University of Cambridge.

Acknowledgements

We would like to thank all of the authors of the chapters of this book who have delivered an eclectic, theoretically informed and above all fascinating set of perspectives on various facets of the human resource management function, the construction labour market and the nature of industry practice. We would like to extend a special acknowledgement to Helen Lingard, co-author of our original text. Although Helen was unable to co-edit this edition with us, she has generously provided a deeply thought-provoking chapter within this volume.

1 HRM in construction: critical perspectives

Andrew Dainty and Martin Loosemore

Introduction

Despite its size and socio-economic significance, the construction sector remains a poorly understood industry, particularly in relation to its people management practices. While industry reports and textbooks alluding to the 'importance' of people abound, too many firms treat people like any other resource to be efficiently managed, or worse to be exploited as 'human capital' in the cause of improved performance (Dainty *et al.* 2007). Given the importance of people in the industry, it is surprising that so little research exists on HRM in the sector. As with HRM practices in other project-based enterprises, there appears to be an assumption that project-oriented firms have specific HRM requirements, and yet research in this area remains limited also (see Huemann *et al.* 2007). Arguably a greater focus on the management of people in construction would better frame debates around management practice and its effects on those who work in the sector.

The management of people within the industry has not been immune to the ubiquitous 'performance improvement' agenda. This movement has arguably tended to subjugate people management as a mere component part of a broader performative agenda. Within the UK, for example, a *Respect for People* working group, itself stemming from the influential *Rethinking Construction* report (Egan, 1998) has produced two significant reports and guidance documents (*Respect for People* working group 2000; 2004). Using critical discourse analysis Ness (2010) reveals how these reports can be drawn upon to legitimise particular arguments, which can result in a further entrenchment of existing power relations just as much as they can improve conditions for workers. As Ness states 'The velvet glove of respect for people covers the iron fist of instrumental rationality', ibid.: 490. Thus, whilst it is important that the role and prominence of people management research is foregrounded in current debates around industry change and development, it is similarly important that the power-effects of such debates are understood in relation to their impact on those who work in the industry.

This book is predicated on the view that a crucial first step in reframing the debate around HRM in the construction sector is to expose both the nature of practice, and the dominant theoretical positions used to understand them, to greater critical scrutiny. Since the 1990s critical management studies (CMS) is a label that has been attached to work that has questioned elements of managerial knowledge and practice, including project work (see Cicmil and Hodgson 2006; Alvesson and Willmott 1992; Fournier and Grey 2000). Up until relatively recently, there has not been a particularly strong tradition of mobilising overtly critical positions on practice within the construction management research field (Ness 2010). Rather, the field has tended to pursue outcomes typically rooted in cost-efficient performativity and 'best practice' panaceas. Interestingly, there seems also to be some reluctance to adopt critical perspectives within the HRM field. Here too, a consensus perspective has maintained a performance focus, thereby avoiding theorising on the socio-political and moral implications of HR practices (Keegan and Boselie 2006). There is arguably a need, therefore, for a more critical discourse around HRM practice within construction.

Another overarching aim of this text is to encourage 'reflective practice'; that is for those reading the perspectives presented here to consider the perspectives offered and to make sense of them within the context of the complex realities of their own professional roles (cf. Schon 1983). The positions mobilised within the ensuing chapters are all very different and many run counter to each other. Several contributions level specific criticisms at the editors' earlier book on HRM in Construction Projects (Loosemore *et al.* 2003), especially for its prescriptive and normative nature. In contrast with the earlier edition of this book, our authors do not lay claim to having found answers to the problems which beset the industry, nor do they suggest that the perspectives and issues discussed will necessary resonate across all industry contexts and organisations. Rather, the value of their more critical approaches is in the questioning of established managerial orthodoxies which have seemingly done so little to reconcile the needs of those who work in the industry with those who employ them. In order to achieve this, many of the authors draw upon theories from outside of construction to interrogate industry practice. Others explicitly *challenge* the relevance of such theories to the complexities of construction work practices. Thus, rather than positioning construction as a sector which lags behind others, in this book we seek to develop a deeper understanding of the conditions faced by construction firms, the influences on their strategy which result from such practices and most importantly, what these practices mean for those who work in the industry.

In this chapter we begin by briefly exploring the industry as a context within which to manage people. We explore the reasons as to why theoretical perspectives tend not to resonate with the construction industry context, and we speculate as to what a more critical orientation might offer the industry in rethinking the ways in which people are managed. We then review briefly the contributions of the chapters within the book to highlight some of the debates to which they contribute. All of the chapters relate in some way to practice, with many containing case examples taken from industry. The intention here is to contribute to the broader 'practice turn' in organisational and management studies, a turn which arguably has specific relevance and importance in construction management (Bresnen 2007). We also draw inspiration from the example set by Smith (2007) in relation to sense-making in projects (cf. Weick 1995), in that our authors seek to make sense of people management practices in construction through an understanding of how they are experienced. It is then up to the reflective practitioner to make sense of these contributions within the context of their own understanding and experience.

The construction organisation in context: a problematic arena for effective HRM practice?

Virtually all of the chapters within this book discuss the nature of the industry's structure and implications for the ways in which people are employed and managed. Key concerns in this regard relate to defining the sector, its size and structure, all of which have direct implications for the ways in which firms operate the HRM function. These are explored and problematised by Ness and Green in Chapter 2, but it is worth highlighting some of these salient features in terms of how they relate to the later contributions.

Although highly exposed to the vulnerabilities of economic cycles of boom and bust (see Dainty and Chan 2011), construction output is set to grow rapidly over the next few years. According to Global Construction 2020 (2011) the global industry is set to grow to $12 trillion by 2020, an increase of almost 70 per cent. However, defining the 'construction industry' is especially problematic given its complex and multifarious nature. This problem stems in part from the fact that the industry spans so many different production and service sectors, leading to a set of what can be described as 'narrow' and 'broad' definitions (Pearce 2003). The former excludes many activities that would normally be included within the definition of construction (such as engineering and design services) and so a broader definition which sees construction representing around 10 per cent of gross domestic product is probably more appropriate (see Dainty *et al.* 2007). However, although more accurate in portraying the full spectrum of products and services that it delivers, it also renders it extremely nebulous and complex, especially from a people management perspective. Those working in the industry transcend unskilled, craft and professional occupations, all of whose input must be coalesced within a temporal project-based environment.

Other structural characteristics are important in defining the employment context of the sector. For example, within the UK the deregulated nature of the industry brought about by privatisation and taxation policy has led to an ingrained reliance on large-scale self-employment (Briscoe 1999;

Briscoe *et al.* 2000; Chan *et al.* 2010). This phenomenon has arguably undermined training and skills reproduction within the sector, with larger firms in particular having a declining significance as direct employers (Green *et al.* 2004; Gospel 2010). It is little surprise, therefore, that construction is dominated by small firms who account for the majority of the industry's productive capability (Harvey and Ashworth 1999; Dainty *et al.* 2005).

The outsourcing undertaken by larger firms has had significant implications for both the definition of skills (Dainty and Chan 2011) and the actual employment of labour, which tends to be discarded as levels of demand change. Labour is usually employed contingently through sub-contracting chains (Debrah and Ofori 1997; Forde and Mackenzie 2004; 2007a; 2007b; McKay *et al.* 2006). These can often extend through many layers with profound effects on both the implementation of coherent HRM strategies (Green *et al.* 2004) and for an organisations' ability to control processes for which they no longer have direct responsibility (Grugulis *et al.* 2003). Perhaps more profoundly, this can also be seen to have shaped the 'casual' nature of the employment relationship (Forde and MacKenzie 2007a). Employees and employers have little loyalty towards each other, preferring instead to move between employment opportunities as they emerge. Another corollary of the reliance on contingent labour has been the tendency of most larger construction organisations to act as 'flexible firms' (cf. Atkinson 1984). This organisational typology, which was discussed in the first issue of this book (Loosemore *et al.* 2003) and elsewhere (Langford *et al.* 1995; Johnstone and Wilkinson; and Raidén and Sempik – this volume), represents an enduring model of operation, but one which has similarly profound implications for the investment in people and the reproduction of skills. The few direct employees who do remain in such organisations must, of course, provide flexible skills and behaviours if such organisations are to maintain their competitive positions (Lim *et al.* 2011).

It could be expected that employment policies would have addressed the failures which characterise the flaws in the employment context, but the low barriers to entry and the weak regulatory framework which underpins the industry's labour market (see Gospel 2010) militate against this. Indeed, given this structural employment context and the weak employment relationships which emerge from it, it is little wonder that calls for construction to improve its people management practices have been largely ignored, as Dainty *et al.* (2007) state:

> Human resource issues too often lie outside the remit of project managers who neither know nor care about the employment status of many operatives on the project for which they are responsible. What results is an employment relations climate characterised by separation, conflict, informality and a reluctance to embrace change.

Thus, HRM activities are often regarded as marginal activities within construction firms whose focus tends to reside in site-based production activities and not on the broader labour market capacity and capability of the organisation or wider sector. But this lack of focus on innovative people management practice has implications which extend far beyond the industry's ability to reproduce skills, especially as it reinforces the entrenched fragmentation and parochialism which is widely acknowledged to lie at the heart of the problems that it faces (see Latham 1994; Egan 1998; Wolsthenholme 2009). An elevation of the profile of HRM within construction research and practice is arguably long overdue.

Towards a more critical perspective on construction HRM …

It could be expected that the industry landscape discussed above might challenge the relevance and applicability of mainstream HRM theories to construction organisations. However, very few commentaries on HRM within the sector have challenged or problematised mainstream theories in relation to the operating context that such organisations confront. As Huemann et al. (1997) point out in their review of HR research on project-based environments, previous texts (e.g. Langford *et al.* 1995; Loosemore *et al.* 2003) have tended to apply standard HRM thinking to the industry, rather than exploring whether the industry needs a different approach. This has several implications which form key foci for this book.

First, much mainstream HR theory is fairly normative in orientation and tends to prescribe actions which are geared around performance outcomes. Known as a 'best practice' orientation, this literature suggests that there are certain approaches which will enable companies to achieve competitive advantage (Torrington *et al.* 2008: 21). A more critical perspective on HRM emerged in debates within the UK literature in the 1990s, where the inherent contradictions between HRM models and the rhetorical nature of the discourse were debated (see Gill 2007). However, this criticality has seemingly failed to pervade the literature around HRM in construction. Even recent perspectives on managing and deploying people have grouped human resource inputs into the 'personnel factor' (Belout and Gauvreau 2004), or have focused on the cost implications of labour which is effectively treated as another factor of production (Lin 2011). Whilst such perspectives are certainly valid from a performative perspective, it could be argued that there is a need to balance this debate with perspectives which tackle HRM from an ethical standpoint. Rather than see people as resources to be optimally deployed to production tasks, this sees employees as possessing special attributes which need to be harnessed, nurtured, developed and understood.

A second implication concerns the characteristic of much of the writing on HRM in construction (as well as project-based environments more generally) to focus on how people management can be used to enact improvements in project and organisational performance. The relationship between HRM practices and performance lacks theoretical support, even within the HRM field (Fleetwood and Hesketh 2007). Within the project management field, doubts have been cast as to the potential of HRM (or 'personnel factors') to influence project outcomes (Pinto and Prescott 1988; Belout and Gauvreau, 2004). Thus, it is important that practices that claim to improve performance and productivity are questioned, and that the implications of new business processes are evaluated for the people that work in the industry. This is not to diminish the need for HRM practice to contribute to broader business objectives, but to emphasise the concurrent need to consider the broader effects of HRM practice. A good example of this is provided by Green (1998, 2002) who questioned the implications of lean processes which have been enacted as part of the performance agenda stemming from the Egan Report (Egan 1998). His analysis reveals how such practices, and their tendency to mobilise 'machine metaphors' in pursuit of performance outcomes, could have negative consequences for those expected to deliver them. It is essential, therefore, that the unintended consequences of such practices are better understood, and factored into decisions of how to enact them.

A third need for a critical perspective on HRM in the industry rests on the need for fresh theories which account for the unique circumstances of the construction firm. It is tempting, especially given the shortcomings in our understandings, to simply examine construction HRM practice through a range of established theories of HRM. In reality, however, it is far from certain that such ideas – many of which have been derived from much more stable production and service environments – will resonate with the uniquely complex and fluid environment that the industry provides. This is particularly the case in relation to establishing an employment environment in which people are managed in an ethically responsible manner (see Huemann *et al.* 2007). What seems clear is that the unique context of the industry renders the applicability of such a theory questionable, at least without a proper recognition of the ways in which context shapes it (Dainty and Chan 2011). Extant theories should not be taken uncritically without thought to their applicability in a project-based setting. Indeed, construction offers an ideal arena within which to test the robustness and generalisability of existing HRM theory given its complex and highly dynamic nature.

One way in which a critical orientation could contribute to addressing these apparent failings in extant perspectives is to encourage project-based organisations to break out of managerialist and prescriptive agendas to open up a concurrent trajectory of work which focuses on the individual perspective (Huemann *et al.* 2007). All too often the perspective adopted by researchers reflects those with positional power within construction organisations. This arguably reinforces asymmetrical power relations and tells us little about the actualities of people management practice.

As was alluded to above, a key debate within HR, as it is within other management fields, concerns the issue of whether 'best practice' or a 'best fit' approach should be adopted when it comes to defining the practices which might enable a better enactment of the HRM function (Torrington *et.al.* 2008: 21). Those advocating a best practice perspective see there being clear deterministic relationships between specific HRM practices and performance outcomes. By implication they also suggest that there are more effective ways to manage. In contrast, the best fit perspective takes a more contingent view on practice which recognises that there is no, one, best solution. A recent contribution in terms of how to enact such practices is provided by Delbridge *et al.* (2006: 139) in the form of 'promising practices', where new ideas are effectively appropriated in context by embedding, sustaining and renewing them. This demands capturing not just the abstracted practices, but insights into their situated context if they are to be better understood.

Another potential contribution of critical perspectives is to better understand the informal and emergent nature of practice within many construction firms, rather than seek to characterise, rigidly define or formalise it. In recent years there has been a 'practice turn' within organisation and management studies (see Nicolini *et al.,* 2003; Whittington 2006; Bresnen 2009). In this perspective, rather than 'black-boxing' industry practice within models or organisational stereotypes, there has been a conscious effort to better understand practice. Such work has sought to contribute to developing 'theory of practice' as opposed to 'theory for practice' (cf. Cicmil *et al.* 2006). To this end, most of the positions on industry practice presented within this book are rooted in, or at least illustrated by, industry case examples. The authors have sought to shift the focus of the debate away from asking how organisations structure themselves to cope with pressures such as demand fluctuations and the need to maximise cashflow, towards understanding the effects of what they do on the people who work in the industry.

Continuing this theme, authors such as Cicmil (2006); Bellini and Canonico (2008) and Smith (2007) have developed a range of more reflexive and informal perspectives on project management knowledge which underpin a different debate and set of provocations for those involved in project-based forms of organising. In the introduction to their collection of critical perspectives on project management, Cicmil and Hodgson (2006) explain how critical positions provide a wider perspective, a better understanding and insight into '…. what determines the position, agendas and *power* of different participants, and how these different agendas are combined and resolved in the process by which the *decisions* are arrived at' (ibid., 12, emphasis as in original). Their aim is not to suggest that a particular alternate view or form of critical analysis is more appropriate, but, following Alvesson and Willmott (1996), to encourage perspectives from a range of alternative theoretical perspectives in order to counter the instrumental rationality which pervades much of mainstream project management theory and practice. This line of argument has many resonances for the construction management research field, both because the organisation of construction epitomises project-based forms of working (Dainty *et al.* 2007), but also because so much of the writing on people management in construction has relied upon the normative and prescriptive approaches discussed above. These have effectively legitimised the workplace power relations within which the discourse of performance improvement has become so firmly rooted (e.g. Latham, 1994; Egan, 1998). Essentially, a more critical discourse around HRM reveals that a supposed concern for people may mask a harsher reality of asserting managerial control (Gill 2007; Ness 2010). Such a perspective acknowledges both the productive and negative power effects of discourse, and suggests that it may constrain as well as enable action (cf. Foucault, 1977).

New directions in HRM research for construction

This chapter has thus far argued for more critical perspectives on HRM in the industry in order to provide fresh perspectives and provocations for both the research and practice communities. Our point of departure is that the dominant performative and theoretical perspectives within the construction academic literature with regards to people management must be questioned if we are to address the seemingly intractable problems which confront the industry. We have argued for a more critical focus which *questions* rather than accepts existing HRM theory, which focuses on *individuals* rather than the firm, which focuses on *ethics* rather than performance, which is *contingent* rather than deterministic and which focuses on *best fit* rather than best practice. However, in inviting the contributions within this book we have not imposed a definition of what we mean by 'critical', nor have we required authors to ground their perspective within any particular literature or tradition. Rather, we have asked authors to define and convey their own perspective on the issues that they discuss. What unites the chapters is that they all provide a bridge between the research and practice communities by on the one hand, mobilising theory as a lens on construction HRM practice, and on the other by providing an empirical challenge to the ability of mainstream theory to account for the nuances and specificities of construction.

In Chapter 2 Kate Ness and Stuart Green pose a critical and thought-provoking problematisation of HRM as it applies to the construction context. Reflecting many of the following chapters, their position is predicated on the contention that no single interpretation of HRM can account for the multiple contexts which characterise the construction sector. For Ness and Green however, HRM can be seen as a powerful discourse, and HRM practices in construction tend to be framed around debates within this field, rather than issues surrounding the employment of construction workers. Consequently, assertions as to the industry's approach to HRM are disconnected from the actualities of construction work. The reality is that people management practices derive from the fractured structure of the industry, and HRM has been absorbed into the broader performance agenda. They bring this into focus with a discussion around issues such as the hollowed-out nature of construction firms and the 'invisible worker'. This reveals the disconnect between the literature-based perspective of management and the harsher reality of a divided labour market and an industry populated by firms who focus on price, profit and contracts rather than employment relations, labour and the human point of view. Nevertheless, as Ness and Green eloquently state, it is important to recognise that no matter how rhetorical, the discourse of industry change has progressively shaped the reality of working lives in the sector.

In Chapter 3 Linda Clarke, Charlie McGuire and Christine Wall chart the twentieth-century evolution of building labour in Britain, contrasting this with the development in other European contexts. This chapter provides a fascinating historical backcloth to many of the other contributions within this volume by charting the development of the industry's labour market and its relationship with other socio-economic influences. Their chapter reveals distinct phases in the development of building labour which shifts from one of free collective bargaining, through social partnership, state regulation and ultimately to the occupational/qualified labour market that we have today. Each phase has had marked implications for the division of labour, the vocational education and training (VET) system, wage reforms and relations and industrial organisation. Key social, economic, political and technological changes and events have punctuated and shaped this evolution, which has eventually led to marked differences between the situation in Britain and its Western European counterparts. One corollary of this is that many areas are not formally recognised as 'skilled' with a resulting division between craft and general operatives on site. This contrasts markedly with Germany for example, where a social partnership exists with a wage structure based on hours worked and qualifications rather than output. Britain's division of labour has led to a reduction in the depth of training, to the increasing fragmentation of the construction process and ultimately to lower levels of productivity relative to its Western European counterparts. It would seem that many of the problems which beset the UK construction sector can be laid firmly at the feet of VET and the regulation of labour more broadly.

In Chapter 4 Paul Chan and Mick Marchington reveal how changing organisational forms and institutional contexts render deeply problematic the creation of human resource development (HRD) approaches which align with the strategic needs of firms. This provides an interesting counterpoint to Loosemore, Higgon and Aroney's chapter (see below). The disconnect between the performative rhetoric and operational realities of HRD, and the resultant difficulties of realising the alignment of HRD with broader strategic agendas, is revealed through a series of case studies. Here, Chan and Marchington explore the intervening effects of the processes of negotiation between managers and workers, and between organisations in pursuing skills development. The highly networked inter-organisational context of construction work, together with the lack of social partnerships and industrial democracy within the industry, can be seen to firmly push HRD towards the satisfaction of short-term commercial pressures. This situation is further entrenched by the prevailing VET system (described in detail in Chapter 3). The asymmetries of power, lack of responsibility for education and training and weak institutional coordination provide a problematic landscape within which to enact HRD. There is an acute need for transnational learning to take place, so as to promulgate stronger institutional structures that emphasise the importance of skills and knowledge for the development of the global construction industry.

In Chapter 5 Martin Loosemore, Dave Higgon and Don Aroney review the role of human resources in achieving competitive advantage for construction firms. However, rather than root their perspective in relation to improvements in bottom line performance, they focus on the role of the HRM function in shaping the organisation's unique identity. The point of departure for this chapter is to critique traditional top-down notions of strategy. These are problematised for their inability to reflect the complex and emergent nature of strategy in practice. As an alternative conception they explore the application of the Resource Based View (RBV) to strategy development, which suggests that the resources most valuable to an organisation are those which are rare, non-substitutable, imperfectly imitable and imperfectly mobile. They argue that connecting HRM with strategy in this way serves many agendas, especially because the RBV allows for a better alignment between the goals of the organisation and those who work for it, as well as external stakeholders. It can therefore be seen as important for raising the prominence of other agendas such as enacting corporate social responsibility (CSR) within such firms. However, a paradigm shift will be required for many construction firms since success can no longer be measured in simple financial terms, but in broader ways which reflect the identity of the organisation and the ways in which it operates.

In Chapter 6 Helen Lingard provides a deeply thought-provoking contribution to ongoing debates in occupational health and safety (OHS) practice grounded in the structure of the labour market and the harsh realities of working life in the industry articulated in other chapters. However, rather than focusing exclusively on the macro level structure and culture of working practice, she introduces a second level of analysis by unpacking the role of managers in shaping the safety climate. This is particularly important, she argues, because of the decentralised and non-routine nature of construction work. Lingard reveals the importance of developing understanding of what causes accidents and health issues in construction, as well as the need to better understand the social context of work and how it impinges on the personal lives of those working within the industry. She provides a persuasive case for shifting the focus towards the moral and legal obligations of employers rather than financially-motivated arguments. She also posits a *prima facie* case for moving away from control-based forms of HRM in favour of a focus on high-involvement forms of organisation and management, although these approaches diverge from those which currently predominate in the industry.

In Chapter 7 Katherine Sang and Abigail Powell review the structural and cultural impediments to equality, diversity and inclusion (EDI) and work–life balance (WLB) in the sector. They provide a persuasive case for embracing both agendas, and suggest that the 'business case' has had some purchase within the industry. In contrast, however, the ethical case for diversity and equality has had little impact. Their analysis reveals how, despite a number of initiatives aimed at diversifying the sector, its labour force has remained homogenous and a very difficult one within which to balance work and family lives. This situation appears to be rooted in factors as diverse as discriminatory cultures, outmoded procurement practices and informal networks. However, in opening up discussion around the interplay of how these factors shape the exclusionary nature of construction, Sang and Powell open up new intellectual space and research trajectories which might begin to challenge some of these seemingly intractable characteristics of the way in which the sector operates. A series of mutually supporting organisational-level recommendations are provided to help begin to address these issues.

In Chapter 8 Stewart Johnstone and Adrian Wilkinson review the conceptual underpinnings of employment relations as it applies to the construction industry. As they explain, employment relations represents a field of study in its own right as well as a component of HRM. They explain different dimensions of the employment relationship and the influences that stabilise such relations within organisations. However, given the propensity for construction organisations to sub-contract and to use self-employed labour this shifts the debate around how they sustain positive relationships across organisational boundaries. Indeed, the nature of the employment relationship is very difficult to define within such arrangements. Managing a flexible workforce which is largely devoid of trade union involvement is likely to require a different role for both management and the state in maintaining employee relations. Johnstone and Wilkinson raise a set of important debates within the employment relations field around the role of trade unions, employee voice and engagement. How the industry manages such issues in the future will to some extent determine how employment relations play out in the future.

In Chapter 9 Ani Raidén and Anne Sempik unpack some of the challenges inherent in resourcing construction organisations, and seek to reconcile notions of 'best practice' in employee resourcing with the need to develop fair and consistent practices. People resourcing comprises management activities concerned with staffing, performance management and HR administration. Overarching these operational concerns are the strategic drivers of flexible employment, employee engagement and change management, each of which is identified as shaping (and being shaped by) the operational elements. Although resourcing functions might be considered the more prosaic aspects of the HR function, they prove exceptionally complex within the fluid environment provided by the construction industry. A key issue emerges from Raidén and Sempik's analysis which resonates across the other chapters within this volume – the need to balance 'best practice' approaches with those which reflect the localised complexities and micro-political climate of the organisation. More specifically with regards to the resourcing context, the challenge is to simultaneously provide managers with the autonomy to respond to local conditions whilst providing them with a robust framework to ensure the effectiveness of those key HR practices. It is wrong, they argue, to assume that line managers are equipped with the expertise necessary for the effective management of these functions.

Finally, in Chapter 10 Jan Druker explores the crucial and yet neglected area of reward management. She initially roots her analysis in the 'new pay' literature, seeing reward as needing to be integrated with other components of organisation design. However, given the structure of construction operations outlined elsewhere within this book (see e.g. Chapters 1, 2 and 4) she also points to the need to look beyond the single organisation in order to consider reward in the context of the inter-organisational relationships that constitute the project environment. Here, Druker mobilises segmentation theory to explain how construction employers externalise activities beyond the boundaries of their own organisation. Such a perspective brings into play the role of sub-contractors and the self-employed, both highly prevalent in the industry, as well as informal working practices. Indeed, whilst an analysis of published data wage rates reveals a widening pay differential between managers and direct operatives, the situation is clouded by the informal economy. Druker's thought-provoking analysis reveals how little we know about the ways in which rewards are managed and how they influence skills reproduction and the employment relationship. She points to the acute need for research studies to further unpack these issues within the industry.

Conclusions: bridging divides through reflective practice

In this chapter we have attempted to reveal the importance of addressing two essential problems in the study of human resource management in the construction sector. The first concerns the need to address the chasm between theory and practice. So much of what characterises the operation of the sector renders the application of HRM theory extremely difficult at best, or at worst counterproductive. It is essential to reconsider the relevance and applicability of HRM theories which have often been developed within more structured production environments or service sectors. Construction firms' tendency to outsource so much of their productive capability, coupled to the prevalence of self-employment, means that more ordered forms of HR organisation sit uncomfortably within the realities of construction organisation and practice. Here, a broader acknowledgement of the complex and emergent nature of practice (cf. Chan and Raisanen 2009) as well as informality of HR systems (Lockyer and Scholarios 2007) might help to begin to account for the complex industry landscape within which HRM plays out. Moreover, it might also help researchers and practitioners to better understand how knowledge and practice intertwine in construction practice (see Gheradi and Nicolini 2002). As Chan and Raisanen (2009) suggest, the formalisation and systematisation of practices effectively subjugates much of the tacit knowledge inherent within such processes which can be so valuable in affording learning opportunities. The insights provided within this text suggest that imposing rigid theoretical frameworks and normative performative prescriptions are likely to stymie the achievement of performative concerns as much as they achieve them.

A second problematic concerns the divide between the research communities relevant to this debate. This volume is unusual in that it coalesces contributions from researchers from the construction field with those from business and management schools. It is notable how few papers have been published within leading HRM journals to have explored the nature of construction HR practice. Similarly, it is seldom that the key debates within HRM and associated fields have impacted within the construction research community. Addressing the lacuna between the perspectives of these communities would seem to offer an opportunity to begin to reframe the problems which clearly beset the industry.

Rather than see this text as an affront to practice, we suggest that construction firms and HR practitioners should regard the challenges posed by the contributors as a set of 'provocations for action'. As Keegan and Boselie (2006) argue, it is important that HR research challenges dominant discourse in HR practice, drawing attention to new innovative ideas and practices which can improve employee wellbeing. We hope that together, these chapters promote a re-evaluation of both the role and enactment of HRM within the industry. It is hoped that by opening up a broader debate around how alternative perspectives might challenge existing orthodoxies, these contributions might provoke a fresh debate on HRM within the sector, together with a new set of research trajectories. The contributions also strengthen our knowledge and understanding of the lived realities of those working within construction, which provides a crucial first step on this journey.

References

Alvesson, M. and Willmott, H. (1992) *Critical Management Studies*, Sage, London.

Alvesson, M. and Willmott, H. (1996) Making Sense of Management: A critical introduction, Sage, London.

Atkinson, J. (1984) Manpower strategies for the flexible organisation, *Personnel Management*, 28–31.

Bellini, E. and Canonico, P. (2008) Knowing communities in project driven organizations: Analysing the strategic impact of socially constructed HRM practices, *International Journal of Project Management*, 26, 44–50.

Belout, A. and Gauvreau, C. (2004) Factors influencing project success: the impact of human resource management, *International Journal of Project Management*, Vol. 22, 1–11.

Bresnen, M. (2007) The practice turn in organisational studies and construction management research, CME 25 Conference: Construction Management and Economics: Past, Present and Future, 1747–56, University of Reading.

Bresnen, M. (2009) Living the dream? Understanding partnering as emergent practice, *Construction Management and Economics*, 27(10), 923–33.

Briscoe, G. (1999) The construction sector in *Review of the Economy and Employment 1998/9,* Institute for Employment Research, University of Warwick, Chapter 2, 31–40.

Briscoe, G., Dainty, A. R. J. and Millett, S. J. (2000) The impact of the tax system on self-employment in the British construction industry, *International Journal of Manpower*, Vol. 21, No. 8, 596–613.

Chan, P. and Raisanen, C. (2009), Editorial: Informality and emergence in construction, *Construction Management and Economics*, Vol. 27, 907–12.

Chan, P., Clarke, L. and Dainty, A. (2010) The dynamics of migrant employment in construction: Can supply of skilled labour ever match demand?, Ruhs, M. and Anderson, B. (eds), *Who Needs Migrant Workers?* Labour shortages, immigration and public policy, Oxford University Press, Oxford, 225–54.

Cicmil, S. (2006), Understanding project management practice through interpretative and critical research perspectives, *Project Management Journal* 37(2), 27–37.

Cicmil, S. and Hodgson, D. (2006) Making projects critical: an introduction, Hodgson, D. and Cimil. S., *Making Projects Critical*, Palgrave Macmillan, Hampshire, 1–28.

Cicmil, S., Williams, T., Tomas, J. and Hodgson, D. (2006) Rethinking project management: Researching the actuality of projects, *International Journal of Project Management*, 24, 675–86.

Dainty, A. R. J., Ison, S. G. and Briscoe, G. H. (2005) The construction labour market skills crisis: The perspective of small-medium sized firms, *Construction Management and Economics*, Vol. 23, No. 4, 387–98.

Dainty, A. R. J., Green, S. D. and Bagilhole, B. M. (2007) People and culture in construction: Contexts and challenges, Dainty, A., Green, S. and Bagilhole, B. (eds), *People and Culture in Construction: A Reader,* Taylor and Francis, Oxon, 3–25.

Dainty, A. R. J. and Chan, P. (2011) Human Resource Development: rhetoric, reality and opportunities, Chinowsky, P. and Songer, A. (eds), Organization and Management in Construction, Spon Press, Oxon.

Dainty, A. R. J., Grugulis, I. and Langford, D. (2007) Understanding construction employment: the need for a fresh research agenda, *Personnel Review*, Vol. 36, No. 4, 501–08.

Debrah, Y. A. and Ofori, G. (1997) Flexibility, labour subcontracting and HRM in the construction industry in Singapore: can the system be refined? *The International Journal of Human Resource Management*, Vol. 8, No. 5, 690–709.

Delbridge, R., Gratton, L., Johnson, G. and the AIM Fellows, The Exceptional Manager: Making the difference, Oxford, London.

Egan, J. (1998), *Rethinking Construction,* Department of the Environment, Transport and the Regions, HMSO, London.

Fleetwood, S. and Hesketh, A. (2007) Theorising under-theorisation in research on the HRM-performance link, *Personnel Review*, Vol. 32, No. 2, 126–44.

Forde, C. and MacKenzie, R. (2004) Cementing skills: training and labour use in UK construction, Human Resource Management Journal, Vol.14, No. 3, 74–88.

Forde, R. and MacKenzie, R. (2007a) Concrete Solutions? Recruitment difficulties and casualization in the UK construction industry, Dainty, A., Green, S. and Bagilhole, B. (eds), *People and Culture in Construction: A Reader,* Taylor and Francis, Oxon, 26–39.

Forde, C. and MacKenzie, R. (2007b) Getting the Mix Right? The use of labour contract alternatives in UK construction, *Personnel Review*, 36(4), 549–63.

Foucault, M. (1977) *Discipline and Punish: The birth of the prison*, Pantheon Books, London.

Fournier, V. and Grey, C. (2000) At the critical moment: conditions and prospects for critical management studies, *Human Relations*, Vol. 53, No. 1, 7–32.

Gheradi, S. and Nicolini, D. (2002) Learning the Trade: a culture of safety practice, Organization, 9, 191–223.

Gill, C. (2007) *A Review of the Critical Perspective on Human Resource Management*, Melbourne Business School. Available at http://worksbepress.com/carol_gill/11.

Global Construction 2020 (2011). Available at www.globalconstruction2020.com (accessed March 2011).

Gospel, H. (2010) The dynamics of migrant employment in construction: A commentary, Ruhs, M. and Anderson, B. (eds), *Who Needs Migrant Workers? Labour shortages, immigration and public policy*, Oxford University Press, Oxford, 256–8.

Green, S. D. (1998) The technocratic totalitarianism of construction process improvement: a critical perspective, *Engineering, Construction and Architectural Management*, 5(4), 376–86.

Green, S. D. (2002) The human resource management implications of lean construction: critical perspectives and conceptual chasms, *Journal of Construction Research*, 3(1), 147–66.

Green, S., Newcombe, R., Fernie, S. and Weller, S. (2004) *Learning across Business Sectors: Knowledge sharing between aerospace and construction,* University of Reading, Berks.

Grugulis, I., Vincent, S. and Hebson, G. (2003) The rise of the 'network organisation' and the decline of discretion, *Human Resource Management Journal*, 13(2), 45–59.

Harvey, R. C. and Ashworth, A. (1999) *The Construction Industry of Great Britain*, Butterworth Heinemann, Oxford.

Huemann, M., Keegan, A. and Turner, J. R. (2007) Human resource management in the project-oriented company: A review, *International Journal of Project Management*, Vol. 25, 315–23.

Keegan, A. and Boselie, P. (2006) The lack of impact of dissensus inspired analysis on developments in the field of human resource management, *Journal of Management Studies*, 43(7), 1491–511.

Langford, D., Hancock, M. R., Fellows, R. and Gale, A. W. (1995) *Human Resources Management in Construction*, Longman, Essex.

Latham, M. (1994) Constructing the Team, HMSO, London.

Lim, B. T. H., Ling, F. Y. Y., Ibbs, W. Raphael, B. and Ofori, G. (2011) Empirical analysis of the determinants of organizational flexibility in the construction business, *Journal of Construction Engineering and Management*, Vol. 137, No. 3, 225–37.

Lin, K-L. (2011) Human resource allocation for remote construction projects, *Journal of Management in Engineering*, Vol. 27, No. 1, 13–20.

Loosemore, M., Dainty, A. R. J. and Lingard, H. (2003) HRM in Construction Projects: Strategic and operational approaches, Taylor and Francis, Oxon.

Lockyer, C. and Scholarios, D. (2007) The 'Rain Dance' of Selection in Construction: rationality as ritual and the logic of informality, Personnel Review, Vol. 36, No. 4, 528–48.

McKay, S., Craw, M. and Chopra, D. (2006) *Migrant Workers in England and Wales: An assessment of migrant worker health and safety risks* (London, HSE).

Nicolini, D. Gheraldi, S. and Yanow, D. (eds; 2003) *Knowing in Organizations: A practice-based approach*, M. E. Sharpe, New York.

Ness, K. (2010) The discourse of 'Respect for People' in UK construction, *Construction Management and Economics*, 28(5), 481–93.

Pearce, D. (2003) *The Social and Economic Value of Construction: the construction industry's contribution to sustainable development*, nCRISP, Davis Langdon Consultancy, London.

Pinto, J. K. and Prescott, J. E. (1988) Variations in critical success factors over the stages in the project life cycle, *Journal of Management*, Vol. 14, No. 1, 5–18.

Respect for People Working Group (2000) *A Commitment to People 'Our Biggest Asset'*, Rethinking Construction.

Respect for People Working Group (2004) *Respect for People: A Framework for Action*, Constructing Excellence.

Schon, D. (1983) The Reflective Practitioner: How professionals think in action, Basic Books, New York.

Smith, C. (2007) Making Sense of Project Realities, Gower, Aldershot.

Torrington, D., Hall, L. and Taylor, S. (2008) *Human Resource Management* (7th edn), Pearson, Essex.

Weick, K. (1995) *Sensemaking in Organizations*, Sage: London.

Whittington, R. (2006) Completing the practice turn in strategy research, *Organization Studies,* Vol. 27, No. 5, 613–34.

Wolstenholme, A. (2009) *Never Waste a Good Crisis*, London, Constructing Excellence, available online at: www.constructingexcellence.org.uk/pdf/ Wolstenholme_Report_Oct_2009.pdf.

2 Human resource management in the construction context: disappearing workers in the UK

Kate Ness and Stuart Green

Introduction

There are two main challenges involved in writing about human resource management (HRM) in the construction sector. The first relates to the definition of HRM, and the second relates to the definition of the construction sector. The purpose of this chapter is to explore the lack of connectivity which often characterises these two ill-defined constructs. It is important at the outset to emphasise that the construction sector is not a homogeneous entity. Issues of boundary definition invariably precede any discussion of construction's supposed defining characteristics. The UK construction sector employs more than two million people and comprises approximately 8.5 per cent of GDP. If a broader definition is adopted to include the construction value chain in its entirety (i.e. architecture, engineering and construction products) the respective figures rise to three million and 10 per cent (LEK, 2009). But the sector has an additional crucial role in supporting the competitiveness of UK industry as a whole. Its capacity has long-since been central to government policy objectives in health, education and transport. Construction further sits at the nexus of society's aspirations for social, economic and environmental sustainability. It is further evident that the construction sector comprises an increasingly popular empirical context for scholars from business schools. The difficulty is that mainstream management theories often have little purchase in the fractured terrain of the construction sector.

It is frequently stated that the construction sector is highly fragmented, with low barriers to entry and many small firms. In 2009 there were in excess of 200,000 construction companies recorded in the official statistics (ONS, 2010), although such statistics about the construction sector should routinely be treated with suspicion (Briscoe, 2006; Cannon, 1994). But what is clear is that it is not possible to understand HRM in the construction sector without also understanding the contracting system. Main contractors typically manage projects, but do not actually carry them out. They have over time become increasingly divorced from the physical act of construction. Many projects are characterised by long chains of sub-contracting culminating in a workforce a significant proportion of which is notionally self-employed. The industry is further characterised by low margins, competitive tendering, and a managerial emphasis on risk avoidance. One of the most notable characteristics of the UK construction industry is its ability to expand and contract in response to severe fluctuations in demand. The Egan Report (DETR, 1998) cites the flexibility of the industry as a key strength. Yet the flexibility of the contracting sector has been achieved at the cost of insecure, casual employment on a project-by-project basis. This is the terrain at which the habitually ill-defined recipes of HRM are directed, and not surprisingly they frequently find little purchase Hence the objective of this chapter is not to espouse the need for ever more sophisticated HRM practices in the construction sector, but rather to problematise a debate which too often lacks meaningful contextualisation.

Understanding construction

The nature of construction work

The evolution of the construction improvement agenda in recent years implies that it is only professional work which has intrinsic interest; innovation is too often assumed to be the preserve of the professional. But construction work in the trades also provides a high level of intrinsic interest with significant scope for innovation. Site work is rarely narrowly defined, repetitive, or closely supervised. Construction work offers freedom from the detailed surveillance of many modern jobs. Working in different places and with different people is the positive side of project-based work. Many find intrinsic satisfaction in physical, creative work where the results are clearly visible at the end of each working day. Innovation is to be found in day-to-day problem-solving and in dealing with unforeseen circumstances. Personal relationships are also important. Camaraderie and a site culture of horseplay and 'having the crack' go along with the feeling of belonging which comes from shared danger and discomfort (Applebaum, 1999). Despite the repeated exhortations of the harbingers of change, collaborative working within work teams is the norm rather than the exception. But collaboration frequently exists in isolation of any managerial imperative (Green, 2006).

But it is also possible to paint too rosy a picture of life on a construction site. Construction workers frequently suffer from job insecurity, long hours, and poor working conditions. Despite the increasingly pervasive discourse of 'modern methods of construction' (MMC), the majority of construction work still takes place out on site, in all weathers. Site facilities were notably described by the Egan report (DETR, 1998) as 'typically appalling'. Workers are routinely taken on and shed in accordance with the season and the business cycle. They are required to re-locate from place-to-place as the work moves around, and move from contractor-to-contractor depending on which one bids successfully for work. Construction is undoubtedly one of the most hazardous industry sectors; construction workers are too frequently killed, maimed or made ill by their work (HSE, 2001; 2003; Jones *et al.*, 2003). One third of all work fatalities happen in construction, and construction workers are six times more likely to be killed at work than employees in other sectors (Donaghy, 2009).

Construction in the UK is also extraordinarily male-dominated. Whilst women now make up 49 per cent of the UK workforce, and have made considerable advances in many fields of work since the passing of the Sex Discrimination Act in 1975, construction remains 'largely impervious' to equal opportunities legislation (Duncan *et al.*, 2002). Only around 10 per cent of construction employees are female and many of these are office-based administrative staff; far fewer professionals and project managers are female, and less than 1 per cent are site operatives (Clarke *et al.*, 2004). Ethnic minorities are also under-represented, in spite of a tradition of successive waves of immigrants working their way up through the construction industry. For some women, ethnic minorities, or others who are 'different', the building site can be a hostile and threatening environment (Fielden *et al.*, 2000; Dainty *et al.*, 2002; EOC, 2005; Gale and Davidson, 2006). Those women or minorities who do persevere tend to be those who 'fit in' with the culture, enjoy the work and become very committed to the industry.

Construction industry improvement and the disappearing worker

In previous decades, issues relating to the workforce were central to the construction improvement agenda. The Emmerson (1962), Banwell (1964) and Wood (1975) reports were consistent in linking improved working conditions with direct employment. Labour casualisation through the 'lump', whereby non-unionised labour is hired on a daily basis and paid cash-in hand, was held to be highly damaging. In contrast, the public sector Direct Labour Organisations (DLOs) were seen to provide exemplary employment contexts which the private sector would do well to emulate. At the time, it was seen to be axiomatic that good conditions equated to direct employment. Much was made of the importance of providing contractors with demand stability and predictability. The latter argument was made especially strongly by Wood (1975); however the Wood Report was rapidly left behind by changes in the political landscape (Ive, 2003). Macroeconomic management very quickly became yesterday's narrative as the enterprise culture was unleashed. Stability and predictability soon became as outdated as the 'social contract'.

The economic environment which prevailed in 1975 when the Wood Report was published was very different from that which prevailed in 1971 when the working party first met. The intervening period saw extensive industrial unrest and economic uncertainty. The winter of 1971–72 saw the first national miners' strike since the 1920s. 1972 saw the first ever national building strike which left a lasting legacy of bitterness. The early 1970s were further characterised by rocketing inflation; unemployment reached one million for the first time since the 1930s. Economic uncertainty was exacerbated by the OPEC oil crisis of 1973. This was quickly followed by a second miners' strike which resulted in phased power cuts and the implementation of the three-day working work. In February 1974 Prime Minister Edward Heath called for a general election on the issue of 'who governs Britain?'. The British electorate responded with a resounding 'not you!', although they were not very convinced about the opposition either. The immediate outcome was that Harold Wilson's Labour Party formed a minority government. A second general election in October 1974 was slightly more decisive and secured Wilson a narrow workable majority. This broader political and economic uncertainty impinged directly upon the construction sector, such that output in 1981 was less than output in 1969. Increased economic uncertainty caused many construction firms to shun direct forms of employment in favour of an increased reliance on sub-contracting. Stability and predictability were replaced by leanness and agility in the marketplace, with dramatic implications for employment practices throughout the sector.

The same logic which prevailed within main contractors, also prevailed within sub-contractors and sub-sub-contractors. The quest for leanness and agility therefore inevitably resulted in the growth of self-employment, despite prevailing concerns about its detrimental side effects. In truth, the concerns of the Emmerson, Banwell and Wood reports about the growth of self-employment were simply outflanked by the onset of the enterprise culture. It was not so much that the argument in favour of direct employment had been lost, but more that the argument was no longer perceived to be relevant. The election of Margaret Thatcher's Conservative Party in 1979 acted to confirm the iconic status of the free market. The freedom of firms to operate in accordance with the demands of the marketplace rapidly became axiomatic.

From the mid-1980s onward there was an observable tendency for the construction improvement agenda to move progressively upstream. Concerns about employment conditions noticeably abated as the focus of attention switched to the extent to which the construction industry kept up-to-date with the latest management techniques (cf. University of Reading, 1988; Latham, 1994; DETR, 1998). The debate echoed that within mainstream manufacturing as total quality management (TQM), business re-engineering (BPR) and lean thinking were successively exhorted as panaceas of industry improvement. The net result was that the construction worker became increasingly marginalised from the debate about construction improvement, with the consequence that debates about HRM became the preserve of the few rather than the many. More recently, the improvement agenda has focused on abstract concepts such as 'value in design', 'systems integration' and 'digital practices'. It is no longer fashionable to raise concerns about the employment conditions of construction operatives; these days we hear only a few marginalised voices.

The onset of the enterprise culture was by no means independent of the rise of human resource management (HRM) (Legge, 2005). In many respects, managerialist improvement recipes such as TQM, BPR and lean thinking were constituent parts of the same discourse. It is interesting to chart the development of HRM in parallel with the restructuring of the construction sector.

Shifting and illusive debates in HRM

Simplistic dichotomies

The advocates of 'human resource management' (HRM) have undoubtedly been influential in encouraging a more strategic orientation towards the management of people. The literature on HRM repeatedly emphasises the need to treat people as a key resource. The espoused aim is to integrate human resources policy into strategic management whilst seeking behavioural commitment to organisational goals (Guest, 1987). There is an established dichotomy in the literature between 'hard' HRM and 'soft' HRM (Truss *et al.*, 1997). The former treats people as a resource to be provided and deployed as necessary to achieve organisational objectives. In contrast, the latter sees people as valued assets who offer a source of competitive advantage (Legge, 2005). In essence, hard HRM comprises a philosophy of 'command and control' whereas soft HRM advocates 'empowerment and commitment'. Such a dichotomy is undoubtedly an over-simplification of a complex field where rhetoric and reality are difficult to separate (Legge, 2005). Many organisations undoubtedly apply elements of both. Companies are also often fond of dressing up hard HRM in a soft rhetoric (Truss *et al.*, 1997). Some organisations apply different elements to different groups of workers, or in different situations, or simply because there is no coherent strategy. Different policies tend to be applied to staff perceived to be 'core' and those perceived to be 'peripheral'. But such simplistic dichotomies frequently conceal more than they reveal.

Espoused policy is frequently more coherent than operational practice, which by necessity tends to be pragmatic and opportunistic (Brewster *et al.*, 1983). In more recent years the rhetoric of soft HRM has increasingly given way to that of high commitment management (Watson, 2004; Legge, 2005), otherwise labelled high performance HR. The concept – despite its persuasive appeal – has been plagued by definitional vagueness from the outset. High performance HR is perhaps best understood as characterised by a number of mutually reinforcing practices (see **Table 2.1**). Whilst the high commitment model is widely accepted as best practice, its level of implementation in UK workplaces has long-since been contested (Wood and de Menezes, 1998).

Throughout the 1990s, models of high performance HR became increasingly conflated with Japanese management practices. Rees *et al.*'s

Table 2.1: Mutually Reinforcing Practices of HRM (Adapted from Legge, 2000)

- trainability and commitment as key criteria in employee recruitment, selection and promotion
- extensive use of systems in communication
- teamworking with flexible job design
- emphasis on training and learning
- involvement in operational decision-making with responsibility
- performance appraisal with tight links to contingent rewards, including promotion
- job security/no compulsory redundancies.

(1996) interpretation of lean production is useful in the way in which it distinguishes between the 'hardware' of production techniques such as just-in-time (JIT) and the 'software' of high commitment HRM practices. The Egan report (DETR, 1998) was responsible for popularising lean thinking in the UK construction sector. But as in other sectors, there was a relentless emphasis on the 'hardware' of the production techniques:

> Lean-thinking presents a powerful and coherent synthesis of the most effective techniques for eliminating waste and delivering sustained improvements in efficiency and quality. (DETR, 1996)

Egan's criticisms of the 'appalling' standard of site accommodation and the sector's poor safety record should be welcomed, but they hardly constituted a commitment to high commitment management. The construction sector of course has a long-established predilection for instrumental management techniques (Bresnen and Marshall, 2001). One example amongst many is the Construction Lean Improvement Programme (CLIP) which focuses on process improvement through the elimination of waste. The advocated approach arguably owes more to the simplistic machine metaphors of business process re-engineering (BPR) than it does to the software of high commitment HRM practices. Deeply embedded narratives of industry improvement are not so easily disrupted by esoteric and ill-defined models of HRM.

The psychological contract

Notwithstanding the above, too many sources erroneously assume that managers are able to exercise independent choice between available HRM models. In practice, management action is constrained by factors such as the size and history of the organisation, prevailing market pressure, the nature of the product (or service), employee behaviour and level of trade union activity (Marchington and Parker, 1990). Different sectors have their own industry recipes, and different countries have their own traditions, institutions and cultural specificities. Different policies also tend to be applied to staff perceived to be 'core' and those perceived to be 'peripheral'. Two current trends particularly affect the way in which HRM is exercised in an organisation: the devolution of HR responsibilities to line managers and the shift from collectivism to individual contracts. The demise of collectivism means individual employees are increasingly expected to be personally responsible for their own career development. Trends towards the individualisation of the employment relationship have focused attention on the 'psychological contract' between the employee and the organisation (Rousseau, 1995). The concept is useful in that it considers the implicit mutual obligations and expectations that exist between the two parties. High commitment HR practices are often correlated with a more positive psychological contract, i.e. both parties feel that the other is maintaining their part of the bargain (Guest, 1999).

Contested meanings

What becomes clear from the above is that there is no single interpretation of HRM which extends across different contexts, or across timeframes. The same argument also applies to particular HRM recipes such as high commitment management, diversity, respect for people and teamwork. Not only is there a continuous struggle in the workplace over the practices which are enacted, but also for control over the *meanings* of particular HRM recipes. Each participant or group within any given organisational context can be seen to strive to dominate the definition of what such terms mean, using them strategically to impose their view on others. The struggle over the meanings of management initiatives is part of a wider power struggle over what counts as work, what are acceptable conditions for doing it, who does what type of work and so on. Clegg *et al.* (2001: 32) describe 'business paradigms', the changing discourses which business people use to legitimate management action, as being 'not only of rhetorical significance but also of practical relevance in the way that businesses are run'. The meaning given to concepts such as work, management, efficiency, quality, innovation and knowledge continuously shapes relationships and procedures in organisational contexts. The ascribed meanings also affect what is considered morally acceptable, although Clegg *et al.* are careful to concede that 'no necessary relation exists between the words and the deeds'.

In any particular organisational context some of those involved may be hoping that genuine and lasting changes can be brought about, whilst others may be merely using the rhetoric of high commitment management as conscious obfuscation. Yet even initiatives which are 'intended' (by some) as 'mere window dressing' may bring about real change, as they provide discursive resources which can be drawn upon (Watson, 1995). The idea of HRM as a discourse is further developed by Hardy *et al.* (2000), who describe how discourse can be mobilised as a strategic resource. Discourse is a resource which can more easily be made use of by the dominant group, but can also be taken up by other individuals and coalitions within organisations. It can variously be drawn on, transformed, and integrated into other discourses. In the same way, company policy statements (on diversity or safety for example) which may be 'meant' as empty rhetoric can be drawn on as a resource by employees. This is one reason why it does not make sense to counter-position 'rhetoric and reality' in a simplistic way.

Notwithstanding the above, there are many elements of HRM which are not the subject of overt struggle. This is because they have become naturalised; they are seen as commonsense by all or almost all the participants and thus not seen as ideological or as representing the position of those with most power. This might include, for example, the use of the terms employment and employee, or the shared storyline that HRM is 'a good thing' for both firms and employees. For instance, it has become routinely accepted within HR circles that the human relations approach to management became dominant because of the limitations of Taylorist scientific management in terms of achieving 'empowerment and commitment'. Arguments from outside this frame of reference are routinely ignored. A good example is provided by Braverman (1974), who argues that the human relations approach did not replace Taylorism; but merely refined and extended it:

> the practitioners of 'human relations' and 'industrial psychology' are the maintenance crew for the human machinery.' (Braverman, 1974: 87).

A similar critical perspective is offered by Watson (1986):

> personnel management is concerned with assisting those who run work organisations to meet their purposes through the obtaining of the work efforts of human beings, the exploitation of those efforts and the dispensing with of those efforts when they are no longer required. Concern may be shown with human welfare, justice or satisfactions, but only insofar as this is necessary for controlling interests to be met and, then, always at least cost.

Such views are of course all too easily dismissed on the basis that they are 'Marxist', thereby immediately removing the need to engage with the arguments presented. But at the same time, discourses have to be positioned against something if they are to continue to be persuasive. From this perspective, Marxist-inspired critiques of HRM are fundamental to the continued credibility of the mainstream.

Disparaging noises from the construction site

Getting the job done in a tough world

In seeking to understand the 'peculiarities' of the construction sector, it is important to recognise that construction project managers are frequently resistant to the very idea of HRM. There is a long-established tradition of devolved responsibility for 'people management' in the construction sector. Traditionally, project managers have always had significant discretion over employment issues such as recruitment, training, and health and safety. Collective bargaining on behalf of manual workers by trade unions has never been strong in the construction sector, and has weakened significantly over the last 25 years. Legge (2005) describes how the discipline of personnel management frequently derived its legitimacy from the strength of the trade unions. Low trade union density in the construction sector therefore translated directly to low prestige of personnel departments. The re-invention of personnel management as 'HRM' did little to alter this lack of legitimacy. The advent of HRM also unfortunately corresponded with a significant increase in reliance on subcontracting and the growth of self-employment. Given the unfolding changes in the employment context, the discourses of supply chain management were understandably much more persuasive than those of HRM.

Green *et al.* (2004) describe how project managers from construction are frequently openly antagonistic about HR practitioners. They saw themselves as 'having to get a job done in a tough world' and viewed HR policies as bureaucratic constraints on their autonomy. Similarly in interviews on the topic of lean construction, one interviewee expressed the view that:

> ...it's a tough industry out there. It's about survival. We really don't have the luxury of investing in namby-pamby HR stuff. We have to get things built.' (Green and May 2005)

Whilst this opinion was at the extreme end of the spectrum, there was little evidence from the interviews conducted of any confidence in 'high-commitment' HR practices, or indeed in HRM practices generally. Loosemore *et al.* (2003) also refer to managers in construction seeing the HRM function as 'a necessary liability that must be tolerated' (p. 37) and 'a nuisance' (p. xii).

More 'namby-pamby HR stuff'

The reference to 'namby-pamby HR stuff' also hints at another aspect. 'Namby-pamby' implies weak, without firm methods or policy, lacking in character and directness – all traits which are associated with negative stereotypes of femininity.

The building site is one of the last bastions of a traditional working class masculinity; one of the few remaining workplaces where there is no obligation to be polite, to be 'nice' to people, to dress smartly, or to be computer-literate. This is an occupational culture which takes pride in physical strength and toughness; even managers are expected to roll their sleeves up and 'muck in' on occasion if they are to be accepted and respected. There are frequent references to the 'muck and bullets' of construction, and a tendency to glorify the dirt and discomfort of work on site. The latter tendency can perhaps be understood as a reaction against wider society's disdainful view of construction work as dirty and unskilled. These discourses are such a powerful part of 'construction culture' that even managers who are not site-based, and female managers, draw on them.

Legge (2005: 61–3) points out that both the practitioners and the recipients of early industrial welfare provision were largely female – men could fight for better conditions, whereas women and children working in factories had to be protected, either by legislation or by the 'lady social workers'. This is also related to the low status of the personnel/HR management function. Asked what they disliked about the work, one personnel manager replied 'It's the attachment of women to it – the welfare image' (Watson, 1977; 92–5). This study also found that personnel specialists are seen (by themselves and others) as buffers or mediators between management and labour. But in the final resort they always side with management – it is the *appearance* of being a neutral mediator that is sought (Watson 1977: 175–7). This explains the common accusation of personnel specialists being two-faced. Watson (2008) argues that the tensions between HR managers and operational managers need to be understood structurally. There are basic contradictions within all industrial capitalist societies which require employers to treat employees as exploitable resources and yet, at the same time, to maintain relationships with them as human beings in order to retain their attachment to the organisation.

Against this background, HR managers are not seen as doing real work, but are perceived as useless 'pen-pushers'. Irrespective of whether its practitioners are male or female, HRM is associated with negative 'feminine' behaviour; it is seen as devious, underhand, insincere and scheming. It is further perceived to comprise a treacherous 'two-faced' strategy which avoids overt conflict and presents a polite, smiling face whilst at the same time stabbing one in the back. This is contrasted with the positive 'masculine' traits of a frank, straightforward, no-nonsense culture espoused by most construction managers. Construction culture values 'calling a spade a spade' (or even a 'bloody shovel'); saying what one thinks to someone's face is held up to be a positive characteristic. It is not polite, subtle or tactful, but neither is it deceitful, hypocritical or manipulative. The perceived sly femininity of HRM is seen as a threat to this direct, forthright culture.

Big hat, no cattle

The commonly reported hostility both to HRM as a concept and to its practitioners is by no means confined to construction, although it does appear to be stronger than in other industries. Operational managers often seem to have a negative or highly sceptical view of the HR function. Legge (2005) cites several studies including Blackler and Brown (1980), Guest (1991) and Keenoy (1990). The antagonism dates back to earlier culture change initiatives within the tradition of organisational development (OD). These are variously referred to as 'devious manipulation', 'propaganda' and 'brainwashing'; but also seen as 'wishy-washy', 'being nice' and not to be taken seriously – hence 'big hat, no cattle'. Some managers also complain of 'dictatorial intrusion' into how they manage their own staff (Watson 1986:205).

As already alluded to, one possible reason for construction managers' antagonistic view is that site managers are used to having considerable autonomy to 'run their own show'. Hence they do not appreciate what they see as interference, especially from those with no direct experience of construction. One tenet of construction culture is that 'our industry is unique' and that it is necessary to have experience of it in order to understand it and its people. The idea that specific 'people-management skills' are needed and that HR specialists are those who have those skills is by no means accepted. Most site managers will prefer to engage workers they have previously worked with, or who are recommended by others ('I know Brendan; he's a good lad' (Ness, 2010a)). Failing that, they inspect their tools, or fall back on an unofficial trial period. If HR managers become involved, they will use 'objective' criteria such as qualifications, interviews, or formal references. However, these are generally seen by practicing construction managers as irrelevant to work in the construction trades, and there is strong resistance to the imposition of more bureaucratic criteria.

HR managers are hence often seen as representing a bureaucratic, rule-bound culture which encroaches on the construction site from all sides; they are perceived as part of an encroaching modernity. Many people in construction seem to have a visceral objection to the way in which HRM is seen to 'put people in boxes' by classifying them. The traditional ways of knowing (and of managing) in construction, are firmly embedded in context. In contrast, HRM too often claims universal applicability. Like the earlier invention of management itself it aspires to 'detach a general technology of control from other, more specific and grounded ideas' (Parker, 2002). Advocates of HRM continuously strive for a codified form of knowledge that can be applied across a huge variety of domains, and others continuously strive to undermine such claims.

Surveys from a distance

Hidden subtexts

In light of the preceding section, it is notable that the regularly cited surveys of HRM practice in construction tend to be framed in terms of the debates within HRM, rather than the debates which characterise the working lives of construction workers. Surveys conducted in these terms seem naïve when compared to what might be characterised as the 'real issues'. The hidden subtext beyond many such surveys is to advocate a more 'enlightened' approach to HRM, rather than challenge the relevance of HRM *per se*. Several previous studies contend that the dominant culture of the construction sector consistently emphasises the hard model of HRM. For example, the *1998 Work Employee Relations Survey (WERS)* (Cully *et al.* 1999) investigated three measures of employee participation across 12 sectors: (i) non-managerial participation in problem-solving groups, (ii) operation of suggestion schemes and (iii) formal survey of employee attitudes during the last five years. Such questions are probably meaningful in many industry sectors, and certainly make sense within a typical manufacturing context. But to the vast majority of construction operatives, they would be seen to be of limited relevance if the aim were to understand practice at operative level. The majority of construction workers on major sites have little contact with 'management' from within the main contractor. Many workers do not know, or even care, who the main contractor is. The reported finding that in the construction industry, participation in problem-solving groups occurs in only 21 per cent of workplaces is clearly nonsense, and is an entirely meaningless statistic derived in isolation of any grounded understanding of the nature of construction work.

Participation and influence

The lack of participation in formal problem-solving groups does not necessarily mean a lack of influence over the work. It can be argued that formal problem-solving groups are unnecessary in construction, as it is simply not possible to separate conception and execution of the work to the same extent as in many other occupations. The site-based and highly context-dependent nature of the work, where each job presents unique problems and constantly changing plans, means that many small but complex decisions have to be made every day. Joint problem-solving is the normal way of working. This informal system which makes things work has been described by authors such as Higgin and Jessop (1965) and Applebaum (1982). However, when it comes to issues such as working hours, or site facilities, or the bigger logistical decisions such as siting a tower crane, then the workforce is not consulted. In fact, it is still not unusual for tenders to be priced, and method statements and programmes drawn up, without consulting even the site manager – who then has to work within parameters of time and cost fixed by others.

The WERS question about suggestion schemes must similarly be positioned against the fractured and informal employment contexts which invariably prevail in construction. In truth, construction workers do not have to 'suggest' how things should be done differently; they simply go ahead and do them differently. A constant debate on how things could be done better is characteristic of most construction sites (and is frequently conducted in several languages at the same time). Making suggestions is rarely the problem; more problematic is the capacity or willingness of management to listen to the suggestions being made. Elsewhere, Coffey and Langford (1998) also observed a low level of employee participation in construction, whilst strangely concluding that there are no inherent reasons that prevent effective participation, even at trade level. Such an interpretation would seem to do little justice to the fragmented employment contexts which characterise major construction sites. Perhaps more meaningful are the results of the European survey conducted by Price Waterhouse/Cranfield (Brewster and Hegewisch, 1994) which showed that the status and influence of HRM on corporate decision-making was lower in the UK construction industry than in other European construction industries. Few would challenge that this finding would be very different if the survey were to be repeated in 2012.

Those who seek to advance a more 'enlightened' model of HRM argue that the dominant model of hard HRM sits uncomfortably with the oft-cited need for creative processes and integrated teamwork. The construction sector's supposed regressive HRM culture can also be positioned as detrimental to notions of organisational learning and knowledge management. But the argument that performance would be improved if management enacted soft HRM does little justice to the realities of how work is organised on the typical construction site. It could rather more convincingly be argued that organisational learning and knowledge management are rendered persuasive to main contractors because the need for HRM has been avoided by the widespread reliance on 'supply chain management' (otherwise known as subcontracting).

Rooted in context

Institutionally-embedded practices

Following on from the above is the recognition that HRM practices in the construction sector are heavily mediated by industry structure. Contractors frequently offload the risks and responsibilities of direct employment to subcontractors. Many projects are characterised by articulated chains of sub-contracting that culminate in a workforce that is notionally self-employed. It follows that many contracting firms have no HRM policy for manual employees. HRM practitioners within main contractors are largely confined to dealing with core professional/managerial staff. But it is important to emphasise that industry structure needs to be understood as a dynamic. Static representations of structure are rather less important than an understanding of the dynamics of change. By the same token, it is important to understand HRM not as a defined entity, but as a discourse which is continuously reconfigured over time. More generally, it is appropriate to focus on the extent to which the discourse of HRM has been absorbed into the broader discourse of construction industry improvement. Material changes in the sector's structure can also be considered as material components of the same discourse.

Before further discussing the specificities of HRM in the construction context, it is important to understand the way in which institutionally-embedded practices shape different outcomes. The political, legal, cultural, financial and education/training systems within a particular society work to define the 'rules of the game' within which individual businesses make their choices (Powell and DiMaggio,1991). There are different national business systems in the same way that there are different varieties of capitalism. In particular, there is a clear distinction between the liberal market economies such as the UK and US, and the more co-ordinated economies such as those of Germany, France and the Scandinavian countries (see Clarke *et al.*, 2011 and Chapter 3 of this volume). Bosch and Phillips (2003) distinguish between 'low-road' and 'high-road' strategies for construction industry development. The former are typically adopted within liberal market economies characterised by short-termism in the allocation of capital via stock markets. Liberal market economies are further characterised by lightly-regulated labour markets which tend too easily towards a low-wage, low-skill paradigm. The more co-ordinated (or social market) economies have a greater degree of regulation of capital and labour markets, with a tendency for the banks to provide long-term capital for industry development. According to Bosch and Phillips (2003; 22):

> ... these countries have created the conditions necessary for the creation of industry-specific labour markets in which construction workers develop ties to the industry and construction firms can rely on a stable workforce.

In the context where 'common sense' is framed by the enterprise culture, 'regulation' has almost become a dirty work. The counter-argument is that regulation may well inhibit labour and capital mobility, but through so doing it creates the platform for a high-wage, high-skill and high-quality construction sector. The overriding point of importance is to recognise that the UK construction sector does not need to be the way that it is; choices have been made along the way to follow the 'low-road' route.

The dichotomy between 'low-road' and 'high road' strategies has obvious implications for HRM. In many ways it can be seen to parallel the distinction between the storylines of low commitment and high commitment HRM. It can be argued that the UK business system in general is not conducive to high-trust, high commitment management. It is chronically short-termist with 'an obsession with shareholder value and the bottom line'. UK firms perpetually favour strategies based on mergers and acquisitions rather than organic growth. The business system is further characterised by low levels of investment in R&D, a tendency towards 'lean and mean' forms of work organisation, and price-based competition (Keep and Rainbird 2000). Storey and Sisson (1989) have long since recognised the limitations imposed by de-regulated labour markets and an institutionalised allegiance to short-term cost reduction policies. They further note a deeply ingrained aversion to risk amongst British managers, who are condemned to adopt short-term cost reduction policies due to factors beyond their control.

A project-based industry

On the basis of the preceding diagnosis, the construction sector is not so 'peculiar' after all. But the situation within construction is exacerbated by the project-based nature of the sector and the labour intensive nature of the work. Perhaps most important is the way in which the sector is organised to respond rapidly to fluctuations in demand. 'Jack be nimble, Jack be quick' might well be the motto for the typical construction firm exposed to intensified competition (Bosch and Phillips, 2003). And given the dominant way of organising, it becomes extremely difficult for individual firms to buck the trend and compete on a different basis. Under pressure to win work, even in boom times, contractors struggle to include the costs of training the next generation of construction workers in their tenders while remaining competitive. The same argument also applies to the additional costs of direct employment (see extended discussion below). In essence, contractors compete over which can best escape paying for the industry's long-term needs, to maximise the short-term chances of winning contracts. Labour is seen as a cost to be minimised. The competitive processes at work give firms little choice other than to operate on the basis of the 'low-road' model. This is what institutional theorists refer to as isomorphism (DiMaggio and Powell, 1983), otherwise described as the processes through which firms come to resemble each other.

Construction projects are chaotic, ambiguous, complex and unpredictable multi-party coalitions characterised by 'interdependence and uncertainty' (Crichton, 1966). The factors contributing to this complexity are natural and social, human and technical, internal and external to the project. In particular, interactions between people are complex because the 'temporary multi-organisation' of the construction project brings together participants from many organisations with potentially conflicting priorities. Negotiation and adaptation (and strength of personality!) become more important than ever when the project is a complex coalition of people from different organisations working together. It is not simply a matter of telling (directly employed) 'human resources' what to do; people have to be cajoled, persuaded, bribed or bullied.

Construction projects, with the exception of maintenance work, are by their nature temporary, and must often be delivered to demanding timetables. This in itself can lead to a focus on short-term project goals, tending to exclude the longer-term benefits of 'high-commitment' practices such as training. The product is large and immobile and has to be assembled at the point of consumption, on site. Construction work is hence geographically dispersed; have shovel will travel. This project-based employment environment can create personal and familial financial insecurity and stress (Warhurst *et al.*, 2008). It is frequently necessary for people to travel long distances or to work away from home, making it difficult to achieve a satisfactory work–life balance. For some, constantly moving from project to project, working in new places and with new people, is one of the attractions of the work; but it can also be a source of stress for workers and their families.

Competitive tendering and business cycles

The intrinsic problems for managing human resources which are posed by temporary projects in different locations are exacerbated by the continued (and, indeed, resurgent) prevalence of competitive tendering. Combined with the immobility of the construction product and the exaggerated effects of the business cycle on construction, this makes workloads extremely unpredictable, particularly for smaller firms in local markets. The volume, timing, type and location of work are all uncertain and fluctuating. Construction projects tend to be very large relative to the size of the firms, giving them a 'lumpy' turnover and a high level of risk. This has massive implications for HRM practices. The workers, and sometimes the managers too, must move from firm to firm and from place to place, depending on who successfully bids for work, and may be obliged, in lean times, to leave the industry altogether. Many construction workers are migrants, and this exacerbates the tendency towards a nineteenth century itinerant worker lifestyle, characterised by an endless moving from project to project, long hours of work, poor working conditions, high rates of injury and illness and a reliance on alcohol for relaxation (Fitzgerald 2006; Clarke and Gribling, 2008).

The need for flexibility limits the scope for stable, permanent employment and for workforce planning. When workloads cannot be predicted more than a few months ahead, it is not surprising that strategic approaches to managing human resources seem to be few and far between, and that firms are reluctant to invest in training and development. Yet despite the isomorphic pressures described previously there is some choice as to how the necessary flexibility is achieved. The dominant model is undoubtedly that of numerical flexibility whereby the firm is able to adjust to peaks and troughs in demand through the use of sub-contractors (cf. Atkinson 1984). However, an alternative model – recently championed by Laing O'Rourke – is that of financial flexibility. Under the latter model flexibility is achieved by reducing the basic rates of pay combined with a 'discretionary bonus' which can be deducted in lean times. It should also be noted that few of Laing O'Rourke's employment contracts extend longer than two years, thereby avoiding the statutory requirement for redundancy payments. While UCATT supported the 'shift back to direct employment' championed by Laing O'Rourke, many operatives resisted the new contracts on offer with several wild-cat strikes in response to the pressure to sign up. Other firms have attempted to achieve flexibility through 'multi-skilling' in order to achieve flexibility in the type of work which can be carried out. Amongst traditional tradesmen (sic), multi-skilling is routinely equated with down-skilling and 'chancers' who are willing to try their hand at everything, without being properly skilled in anything (Ness, 2010a).

Flexibility of a different kind

The years 1993 to 2007 comprised the longest period of sustained growth in UK construction since the Second World War. Faced with conditions of labour scarcity, clients talked of partnering and frameworks, and construction firms talked of respect for people. In the manual trades, the 'fortuitous' opening of labour markets to East Europeans helped to prevent wage rates rising, but in the professions, people were in short supply. In the latter half of the 1990s in particular, faced with recruitment difficulties, some construction firms saw a need to improve working conditions and to seek more diverse recruits, rather than simply increasing pay. It was a means of attracting and keeping construction workers under conditions of scarcity. As soon as recruitment became easier – first because of an increase in the supply of skilled workers with EU expansion, then because of a reduction in demand due to recession – most firms no longer saw the need for a 'strategic' approach to HRM[1]. Under pressure from clients to cut tender prices, contractors have certainly not shown much sign of treating people as assets since the recession of 2008.

Nevertheless, there is a paradox here. The complex and unpredictable nature of construction means that work gangs need considerable flexibility and autonomy; management control of location and sequence of work is limited. In a classic article, Stinchcombe (1959) contrasts the 'craft administration' of construction with the bureaucratic administration of mass production. In mass production, elements of the work process such as the location at which a particular task is done, the movement of tools, materials, and workers to this workplace, particular movements to be performed in getting the task done, schedules and time allotments for particular operations, and inspection criteria for particular operations, are centralised and planned in advance by people who do not actually carry out the productive work. In construction, however, they are governed by the worker in accordance with the empirical lore of the craft.

Fifty years after Stinchcombe's article, this still seems to be the case. Felstead *et al.* (2007) studied discretion in work tasks. The question was: 'How much influence do you personally have on ... how hard you work; deciding what tasks you are to do; deciding how you are to do the task; deciding the quality standards to which you work?' When classified by industrial sector, construction had amongst the highest levels of autonomy. When classified by occupation, managers reported the highest levels of task discretion, followed by professionals and skilled trades which were on a level with each other, well ahead of other occupations such as sales, personal service workers, and machine operators. This supports the notion that construction workers at all levels enjoy more control over their work than those in many other occupations. Workers also have considerable influence over the recruitment and selection of co-workers via informal networks (Druker *et al.*, 1996; Clarke and Hermann 2007; Ness 2010b).

However, whilst at the level of the firm there may be an element of 'soft' rhetoric hiding hard reality, on site there may even be an element of the opposite. Site managers need to project a 'tough' image, drawing on 'no-nonsense' discourses of masculinity and apparently treating workers as disposable means to an end. Yet the reality is that power on site is constantly (re)negotiated. In order to get things done, it is necessary to build shifting coalitions, and this is especially true when work is subcontracted – subcontracting and self-employment limit the manager's control. Site managers, and first-line supervisors, often build informal teams of trusted workers who move from job to job together with them, in spite of being notionally self-employed (Chan and Kaka, 2007). The intrinsic satisfaction of construction work and the control over the detailed execution of the work by production workers (Stinchcombe 1959; Applebaum 1999) are also at odds with the supposed need for 'command and control'. All of this tempers the industry's dominant ideology of utilitarian instrumentalism, yet supervisors and managers have limited room for manoeuvre. Even more senior managers often have little scope to depart from established industry recipes. Short-term competitive pressures combine with deeply embedded institutionalised practices to limit the scope for positive change.

Taking the 'low road'...

Behind the curtain of multi-tiered subtracting

The UK construction industry is now characterised by the 'hollowed-out firm' relying on nominally self-employed labour, much of which is supplied through labour agencies or labour-only subcontractors. A survey of building sites in 1995 found as many as five tiers of subcontracting in the chain, with the main project manager having little control over, or even knowledge of, the subcontractors below the second tier. On many sites, 95 per cent of the workforce was self-employed (Harvey, 2001; Harvey, 2003).

Raidén *et al.* (2007) suggest that the concept of multi-tier contracting masks a reality of workers being starved of employment benefits and job security, and feeds the low-cost, low-skill, low-productivity culture which pervades the sector. On some large construction sites with hundreds of workers, each one is supposedly an independent sub-contractor. This manifest absurdity is due to the phenomenon of *false* self-employment, described by the OECD as follows:

> In some countries … taxation systems, and perhaps labour market policies as well, might have encouraged the development of 'false' self-employment – people whose conditions of employment are similar to those of employees, who have no employees themselves and who declare themselves (or are declared) as self-employed simply to reduce tax liabilities or employers' responsibilities. (OECD Employment Outlook 2000: 156, cited in Harvey, 2001)

Such a practice is by no means unique to the UK, being prevalent in many countries, and known in Australia as 'sham contracting'.

Sub-contracting and self-employment have long played an important role in construction, but have increased dramatically since the late 1970s. The current scale of self-employment in the UK construction industry is also exceptional when compared to other countries' construction industries, or to other sectors within the UK. In France self-employment in construction is around 10 per cent; even in deregulated Korea it is only 7 per cent. For the UK workforce as a whole, self-employment is 11 per cent . The self-employed now account for 39 per cent of the construction workforce in the UK, or 57 per cent of operatives (Clarke *et al.*, 2008). Harvey (2001) gives a 'conservative estimate' of the number of *false* self-employed at 361,000 out of an estimated total of around 600,000.

Winch (1998) concludes that the main reason for the growth of self-employment is the strategic choice of construction companies to emphasise flexibility (rather than productivity, or quality) as a source of competitive advantage. However, construction firms do not make a 'strategic choice' in a vacuum; it is shaped by government regulation or lack of it, as well as the institutions and customs of the country. Official UK definitions of employment status confuse *casual* employment – typical of construction – with *self*-employment. Harvey (2001) demonstrates how changes in government policy have determined shifts between direct employment and self-employment over the past 30 years, since self-employment became a simple matter of self-declaration in 1980. More recently, the UK government has allowed entry to 'self-employed' workers from Romania and Bulgaria, whilst refusing to admit the directly employed.

High road comparators ...

Bosch and Phillips (2003) have described how construction industries in the advanced capitalist economies have taken divergent paths in response to the common challenges of competition, volatile demand, and ageing labour forces. Two models can be distinguished. First, the more regulated or co-ordinated 'high road' construction industries, which tend to be 'capital-intensive, human-capital intensive, and technologically dynamic' and characterised by intense state and/or sectoral involvement in training and labour market co-ordination. Second, the low-wage, low-skill, low-technology construction industries on a 'low road' path characterised by extensive subcontracting, 'atypical' forms of employment (temporary and/or agency workers, self-employment, and the informal labour market), and high labour turnover. The customs, institutions and regulatory framework existing in different sectors and countries determine the path taken by the industry.

In the 'high road' construction sectors of Scandinavia, Germany and the Netherlands, qualifications and formal training are essential, wages are set on the basis of collective agreements, and employment is generally direct. In these countries collective bargaining has sector-wide coverage , either due to high trade union membership (as in Denmark) or due to the general application of collective agreements (as in France and the Netherlands). Workers are classified and paid according to their qualifications and experience; in Dutch and Danish construction firms, workers can enjoy substantial job security and climb a hierarchical job ladder (Byrne *et al.*, 2005). The effect of wages, employment conditions, and training being centrally determined – whether by industry bodies or the state – is to take these factors out of the equation when firms are competing for work. By contrast, in Italy, Portugal, Spain, Greece and the UK, which have taken the 'low road' path, labour is employed casually from one project to another, firms are small and self-employment is high. Although official collectively bargained wage rates may exist, in practice earnings tend to be related to output rather than skills and qualifications.

For individuals, 'self-employment' often means no holiday pay, sick pay, unemployment benefit, or pension; no collective bargaining; no employment rights; no training and no opportunity to progress except as an entrepreneur; poor working conditions and poor health and safety provision. It may mean higher income in the short-term, but not over a whole year or a whole economic cycle. It may also mean more control over the work – but in many cases this takes the form of the illusion of being ones own boss – the 'freedom' to work 12-hour days, six or seven days a week.

But there are some advantages

For construction firms, contingent labour gives them the flexibility to respond to changing market conditions by cutting pay rates as well as numbers. It also reduces labour costs, reduces risk, allows them to pay purely according to output and tends to marginalise the trade unions. However, it has adverse implications for training, productivity, and quality. The detrimental consequences of self-employment for training and skill development in construction have been discussed by, amongst others: Forde and MacKenzie, 2007; Winch, 1998; Clarke *et al*, 2008; CITB, 2003; Ball, 1983; Gann and Senker (1998). It reduces firms' ability to organise and manage labour. The multi-layered subcontracting and competition between labour-only gangs makes co-ordination of site work difficult, and 'chaos becomes the norm'.

This situation may not be in firms' best interests, but within the constraints of the wider economic context and (de)regulatory framework, they appear to have little choice. If competitors can win tenders by offloading part of the risks and costs of employment onto workers or society at large, thus reducing labour costs by 20 to 30% (Harvey 2003), then it is difficult for any individual firm to take a longer-term view by employing workers directly and investing in their training or their health and wellbeing. Forms of ownership also condemn many UK companies to a short-term focus, which is often imposed by shareholders. This reinforces the trend towards a low wage, low skill, low productivity, low-tech and low commitment construction industry which externalises many of its costs. The 'choice' of construction companies to emphasise flexibility is of course at odds with a 'commitment to people' HRM approach; it implies treating labour primarily as a cost, and equating competitive advantage with cost-cutting. There is little recognition of people's knowledge, skills, talent and experience as a source of competitive advantage.

Where are those workers?

The workforce all have degrees

> Building, possibly more than any other industry, depends on the abilities, attitudes and adaptability of its workers (BRS, 1967).

By contrast to reports produced in the 1960s on building operatives and their work, in recent policy documents, those who carry out the physical work of construction seem to be increasingly invisible. For example, a report on 'Skills for productivity' (ConstructionSkills, 2006) fails to mention craft skills at all, but concludes that the skills with the greatest impact on productivity are leadership, business management, people management, construction management, and design. Similarly Egan's latest report (ODPM, 2004) concentrates on professional skills in design and management.

DfEE (2000: 10) suggests that it is the increase in subcontracting and self-employment (that 'the site manager may be almost the only employee of the client's primary contractor present on site') which has led construction companies to emphasise skills relating to 'conceiving, scheduling and managing projects'; and to view operational, site-based skills (which do not vary much from company to company) as 'not the real drivers of competitive success or failure'. As most large UK construction firms have become hollowed-out organisations which no longer build anything, the return on capital does not depend on productive efficiency, but on supply-chain management and the manipulation of payments. This seems to be what lies behind the increasing 'invisibility' of the manual workers who make up roughly 80% of the industry. Contractors are progressively more removed from the physical work of construction, choosing instead to concentrate on 'project management'; this is held to be more profitable than actually employing people to build things. Responsibilities for the employment and training of craftsmen are delegated down multi-layered supply chains to small firms.

In *Building* magazine's 'Good Employer Guide 2007', Chrissie Chadney, HR director at Willmott Dixon, is quoted as saying:

> It used to be common for people to come up through the tools into construction management, but that's rare now. The members of our workforce nowadays almost all have degrees and are members of the Chartered Institute of Building or the RICS.

Construction firms these days employ a better class of person, it seems. The workforce 'all have degrees' because much of the workforce, not being directly employed, is exluded from being counted.

Commodified labour

The emergence of human resource management (HRM) and its 'enlightened' policies as something which is done to managers rather than manual workers (Legge, 2005) fits neatly with the denial of responsibility for the actual site operatives who do the work. There seems a curious disconnect between the HRM initiatives and the realities of work on site. Greed (2000, p. 189) comments on:

> the strange contrast between the 'clean' image of management found in the literature ... and the harsh realities of labour relations, characterized by bullying, conflict, exploitation, poor conditions, and pressure ... the 'slave culture' of the building site [is seen] as the 'guilty secret' which professional men are hiding from women seeking to work in construction.

Firms which do not directly employ any operatives very often do not have any training and development policies for the workers engaged by the subcontractors to whom they contract (why should they?). Enlightened HRM policies which apply to 'all our employees' are in fact only for professional, managerial and administrative staff. This has led to the growth of huge disparities in status and reward. The traditional divide between staff and operatives, salaried and hourly paid employees, has been replaced by a division between core (managerial and professional) employees, and workers as commodities, bought in as needed.[2] The engagement of manual workers at arm's length contributes to their invisibility. Labour is even more commodified in this pure market relationship than in the employment relationship; the emphasis on supply chains means that firms are concerned not with employment relations but with price relations; not with labour but with production; and not with the human point of view but the contractual point of view. Labour as a resource is distanced from workers as human beings, enabling many of the moral and legal constraints which had grown up around the employment relationship to be dispensed with. The operative is unlikely to be treated as a person rather than a labour resource, or to be given any but minimum training (Green *et al.*, 2004; Dainty *et al.*, 2007).

Towards a more critical understanding

Enterprise and excellence

The 'enterprise' and 'excellence' literature from outside the construction sector proposes to replace 'outdated', mechanistic, bureaucratic ways of doing business with the flexible, creative, organic, entrepreneurial adhocracy where workers (supposedly) take responsibility for their own work. Workers are to be 'empowered', trained to take on traditional supervisory roles – yet the 'craft administration' thesis (Stinchcombe, 1959) says that construction workers already do this. Oddly, the 'excellence' literature from *within* construction seems to emphasise almost an opposite set of attributes – bureaucratic, Taylorist efficiency, employing 'document controllers' and 'QA managers' and 'HR managers'. Ironically, it seems that, as the construction industry is supposedly trying to become more 'modern', the rest of the world is trying to become 'post-Fordist', flexible, and more like construction. Hence the expression: 'it's a funny old world'.

In some ways, construction is perhaps so pre-modern it is postmodern - a dark version of the bright, flexible future of portfolio careers and individually negotiated contracts. Construction employment was often casual or 'flexible' long before the emergence of the flexible firm model as described by Atkinson (Atkinson 1984; Tressell 1965). Construction firms aspired to be 'lean and mean' well before the emergence of lean construction as a concept in the 1990s. Networks of micro-firms blur the boundaries between workers and entrepreneurs. In construction (apart, perhaps, from the DLOs), the bureaucratic methods of control and co-ordination which were efficient in other sectors never did have much purchase (Stinchcombe, 1959; Applebaum 1982).

Shaping reality

The improvement recipes demand lower costs, higher quality and better productivity. The lower costs at least are often achieved through relentless downward pressure on the supply chain, rather than by improved efficiency of the actual production process – in which most of the big construction companies no longer have any interest. The rhetoric of business process re-engineering or lean construction seems strangely disconnected from the actual processes of pouring concrete or fixing windows, as well as from the people who carry them out. In fact what has gone on is more of a *financial* re-engineering, a reconfiguration of employment and contractual relations. This is demonstrated by a comparison of British and German house-building firms (Clarke and Herrmann, 2004). The German firms studied had their own directly employed operatives and plant. They employed project managers who were building engineers, and emphasised actual construction expertise. In the UK firms, all work was subcontracted; none of the contractors in the sample employed any skilled operatives. The British firms were found to emphasise cost rather than operational expertise; as the authors' remark, it is perhaps no coincidence that the role of the QS seems to have expanded with the increased reliance on subcontracting.

Construction has gone further than most industries in 'the supplanting of bureaucratic principles by market relations' described by du Gay and Salaman (1992). Rather than merely defining internal organisational relations *'as if'* they were customer/supplier relations, where construction operatives are concerned the rhetoric has *become* the reality. This shift has fundamental implications for working practices and for the conduct and identities of workers. It can further be seen as an extreme example of discourse having material effects. The fact that workers are *defined as* self-employed may literally have life-and-death implications (cf. Donahy, 2009). By turning the employment relationship into a contractual relationship, companies have not merely cut costs, but offloaded huge amounts of risk. Responsibility for employment issues is simply passed down the supply chain so that workers, re-imagined as 'independent contractors', become responsible for their own safety. When the price passed down from above via a series of Dutch auctions does not cover the minimum costs of safe access or safe working procedures, main contractors and clients are too easily tempted to adopt a 'don't ask, don't tell' policy, and escape responsibility despite the provisions of the CDM regulations. The contradiction between the demands for ever more work, more quickly, from fewer employees, and the simultaneous moves towards 'respect for people' in the improvement recipes, is thus partly explained by the exclusion of much of the workforce from the 'enlightened' influence of HRM.

Improved organisational performance

The assumption that HRM (especially 'strategic' HRM or high commitment/ high performance HRM) is 'a good thing' for both firms and employees is shared by many, including most 'enlightened' senior managers and policymakers. Perhaps not surprisingly, it underpins most books on the topic, both academic and practitioner-oriented. For example, Loosemore *et al.* (2003) overtly set out to 'persuade project managers of the strong link that exists between HRM performance and project performance' (Loosemore *et al.* 2003: xii).

There is a considerable body of academic work which attempts to show a link between 'enlightened' HR practices (defined in various ways) and the performance of firms (again, defined in various ways). However, a review by Legge (2005: 25–32) casts serious doubt on the validity, reliability and generalisability of much of the data. Most studies have involved large-scale postal surveys of 'single respondents answering quick questions'. Respondents report their assessment of HR practices and strategies across a whole corporation, which may work across different industry sectors and have multiple locations and categories of employee (and therefore varying business strategies). Indicators of organisational performance were mainly financial.

Most studies do not attempt to distinguish between espoused and actual HR practices, nor, perhaps most importantly, to assess the aspects of employee behaviour which are assumed to explain the relationship between HR practices and organisational performance. This means that if a correlation appears to exist between HRM policies and organisational performance, causality is merely assumed, not tested. HRM policies may indeed lead to positive outcomes – though perhaps through some form of Hawthorne effect. On the other hand, as a firm becomes more profitable it may invest more in high commitment HRM practices such as training, so the assumed causality is reversed. Legge also points out the possibility that survey respondents' own theories of the relationship between HR practices and organisational performance are likely to bias their responses. Studies by Gardner *et al.* (1999) and by Guest *et al.* (2003) support this view, finding a strong relationship between HRM and performance when *subjective* measures were used, but not when '*objective*' measures were used.

Culture change

HR specialists are taught to believe that they can manipulate organisational culture using tools such as: selection and promotion processes (controlling who enters the organisation and reaches positions of influence within it); induction and socialisation (influencing social dynamics within the organisation); appraisal and reward systems (rewarding desired traits); and codes of practice etc (influencing employees' work practices) (see Brown, 1995). This is a rather frightening picture, but fortunately it is almost certainly wrong. Culture is not easily subject to conscious manipulation (see, e.g., Willmott 1993; Legge 1994), and construction culture seems to be extraordinarily ingrained and resilient. The industry's culture is also mutually constituted with its structure, so one cannot be changed without the other.

The ubiquitous 'business case' argument

There is an argument, presented by Legge and summarised above, that the relationship between HRM and organisational performance may not be one of straightforward cause and effect. This would undermine the 'business case' often put forward for high commitment management. There is also an argument (see Ness 2010a) that, whatever the nature of the relationship, to argue for better employment practices on the basis of a business case is a dangerous double-edged sword. The business case argument means that improvements to working conditions are judged purely in accordance with their contribution to efficiency and profitability rather than in terms of moral imperatives or fairness, or even compliance with the law. Noon (2007) explores the potential dangers of the business case approach to the employment of ethnic minorities, making the right to fair treatment contingent upon an economic rationale which ignores concerns for social justice or moral legitimacy. There may be cases where it is in firms' economic interest to discriminate. Similarly Davis (2004), points out that there are many situations where safety does *not* pay (because victims, their families and the state bear the costs of injury and ill-health). The economic argument on safety also means the rational employer will prioritise avoiding the accidents that cost the firm most, rather than those with the greatest risks to workers' health and safety.

The business case argument that better employment practices lead to improved organisational performance underpins many policy initiatives such as *Investors in People* or the 'Opportunity 2000' programme to improve women's employment. Of particular relevance to the construction industry is the *Respect for People* agenda, put forward in the reports: *A Commitment to People 'Our Biggest Asset'* (2000); *Respect for People: A Framework for Action* (2004) and *Respect for People: The Business Benefits* (2007) (RfP 2000, RfP 2004 and RfP 2007). The Egan Report (DETR, 1998) had identified 'commitment to people' as one of the essential 'drivers for change' in the construction industry, and the Respect for People working group was set up to look at 'how the construction industry could improve its performance on people issues'. The first report (RfP 2000:6) uses a 'virtuous circle' business case model of 'better pay and conditions for employees' leading to 'happier, healthier and more productive employees' which leads to 'better delivery on quality, cost and time' which leads to 'more satisfied clients and better profitability' which then leads back to 'better pay and conditions for employees'. This would be almost laughable if it did not have such serious implications. The links can be questioned at every stage. There is no empirical evidence; it is merely assumed in order to proselytise for 'better practices'. The report proposes measuring progress by the use of Key Performance Indicators (KPIs). Yet the follow-up report, *Respect for People: The Business Benefits* (2007) does not report on any of these KPIs, it merely states that a survey of those who had used the 'toolkits' agreed that Respect for People contributed a positive business benefit – and goes on to describe unremittingly positive 'case studies' which appear as self-promotion by the firms involved.

However, Pepper *et al.* (2002) tell a different story. Interviews were conducted with representatives of all the companies who had engaged in the trial of the Respect for People 'workplace diversity' toolkit. The business case argument did not appear to be the primary driver behind companies engaging in the diversity toolkit trial. One contractor commented: 'This business argument is fine words with fine goals, but where's the evidence?' Motives for addressing diversity issues varied. Several had misunderstood the meaning ascribed to diversity by the RfP initiative, seeing it as 'how many different things you can do successfully to make money'. Some referred to the need to comply with equal opportunities legislation. Others viewed diversity purely as a marketing tool:

> .... maybe we will score more brownie points if we include some words on diversity, so when a Local Authority client asks for our policy they maybe impressed somewhat ... but I don't believe diversity is an issue within our organisation and it seems a nonsense to have to do it.

In the construction sector, much of what is being promoted as best practice (in initiatives such as RfP) does not comprise 'advanced' initiatives such as employee involvement, but merely involves complying with the basic requirements of employment law. The imperatives rarely stray beyond taking reasonable steps to reduce risks to workers' health and safety, not evading tax by false self-employment and not discriminating against women and ethnic minorities. Whilst the business case may not be compelling, the legal and moral arguments appear much stronger.

Conclusion

It would not be within the spirit of this chapter to offer recommendations on how HRM could be better implemented in the construction sector. Indeed, one of the core arguments developed is that the very language in which debates about HRM are conducted serves to nullify any grounded understanding of the construction context. Too many authors are rather too interested in positioning themselves as harbingers of industry improvement. In truth, the advocates of change are two-a-penny. Rather less common are those who seek to immerse themselves in the intricacies of construction and the way in which the labour market operates. It is also true that the construction sector is attracting increasing attention from mainline HRM scholars as an empirical context. This is a trend which is undoubtedly to be welcomed, but too often too little care is taken to understand the unique dynamics of the construction context. Understanding does not only require a knowledge of how the industry operates in the here-and-now; it also requires an understanding of how the industry has changed over time. Part of this understanding can be gleaned from studying the dynamics of change. These can be theorised in a variety of different ways – for example – on the basis of economics, organisation studies and industrial relations. Yet an important part of any understanding must also embrace the rhetoric of industry change, and the way this has served to progressively shape the reality of working lives in the sector. Enlightened ways of working are unfortunately directly implicated; and human resource management must accept its part of the blame.

But any contextualised understanding must also embrace a knowledge of the construction process, and the technologies and skills which are routinely mobilised – and by whom. It is this last point which has formed the central argument mobilised in the current chapter. The people who work in the construction sector are frequently disconnected from 'management', at least as it is traditionally understood. Operatives in construction are especially disconnected from human resource management; they are beyond its reach and beyond its sphere of relevance. If we dismiss the rather ideologically-loaded lexicon of human resource management and default to 'people management', the conundrum becomes especially focused. The construction sector rather gave up on people management several decades ago. Fragmented employment terrains and post-modernist forms of organisation were developed in construction even before such concepts were recognised as worthwhile arenas of academic study. There is therefore a persuasive argument that the construction sector is leading the way for other sectors to follow. The challenges and pitfalls are there to be seen for those interested enough to observe. But too often the contextualised realities of practices within the construction sector remain stubbornly off the radar for the majority of authors and researchers.

Notes

1 Unless of course the strategy is to take a 'softer' approach in 'tight' labour markets, and a 'hard' approach in recession – which is perfectly possible and logical, if somewhat cynical.
2 However, in major recessions even the core professional and managerial employees turn out to be disposable.

References

Applebaum, H. A. (1982) Construction management: traditional versus bureaucratic methods, *Anthropological Quarterly*, 55(4), 224–34.
Appelbaum, H. A. (1999) *Construction Workers USA*, New York: Greenwood Press.
Atkinson, J. (1984) Manpower strategies for flexible organisations, *Personnel Management*, 16(8), 28–31.
Ball, M. (1988) *Rebuilding Construction*, London: Routledge.

46   *Kate Ness and Stuart Green*

Banwell, Sir Harold (1964) *The Placing and Management of Contracts for Building and Civil Engineering,* London: HMSO.

Blackler, F. H. M. and Brown, C. A. (1980) *Whatever Happened to Shell's New Philosophy of Management?: Lessons for the 1980s from a major socio-technical intervention of the 1960s,* Farnborough: Saxon House.

Bosch, G. and Phillips, P. (eds) (2003) *Building Chaos: An international comparison of deregulation in the construction industry,* London: Routledge.

Braverman, H. (1974) *Labor and Monopoly Capital: The degradation of work in the twentieth century,* New York, Monthly Review Press.

Brewster, C. and Hegewisch, A. (eds) (1994) Policy and Practice in Human Resource Management, London: Routledge.

Brewster, C., Gill, C. and Richbell, S. (1983) Industrial relations policy: a framework for analysis, K. Thurley and S. Wood (eds), *Industrial Relations and Management Strategy,* Cambridge: Cambridge University Press, 67–72.

Bresnen, M. and Marshall, N. (2001) Understanding the diffusion and application of new management ideas in construction, *Engineering, Construction and Architectural Management,* 8(5/6), 335–45.

Brown, A. D. (1995) *Organisational Culture,* London: Pitman.

Building Research Station (1967) *Building Operatives' Work,* London: HMSO.

Briscoe, G. (2006) How useful and reliable are construction statistics? *Building Research & Information,* 34(3), 220–29.

Byrne, J., Clarke, L. and Van der Meer, M. (2005) Gender and ethnic minority exclusion from skilled occupations in construction: a Western European comparison, *Construction Management and Economics,* 23(10), 1025–34.

Cannon, J. (1994) Lies and construction statistics, *Construction Management and Economics,* 12, 307–13.

Chan, P. and Kaka, A. (2007) The impact of workforce integration on productivity, in *People and Culture in Construction: A Reader* (eds A. Dainty, Green, S. and B. Bagilhole), London, Taylor & Francis, 240–57.

CITB (2003) Construction Skills Foresight Report 2003, Norfolk: Construction Industry Training Board.

Clarke, L. and Herrmann, G. (2004) Cost vs production: disparities in social housing construction in Britain and Germany, *Construction Management and Economics,* 22(5), 521–32.

Clarke, L., Pedersen, E. F., Michielsens, E., Susman, B. and Wall, C. (eds) (2004) *Women in Construction,* Brussels, European Institute for Construction Labour Research.

Clarke, L. and Herrmann, G. (2007) *Divergent divisions of construction labour: Britain and Germany,* Dainty, A., Green, S. and Bagilhole, B., (eds) People and Culture in Construction: A Reader, Taylor & Francis, London.

Clarke, L. and Gribling, M. (2008) Obstacles to diversity in construction: the example of Heathrow Terminal 5, *Construction Management and Economics* 26(10), 1055–65.

Clarke, L., McGuire, C. and Wall, C. (2011) The development of building labour in Britain in the twentieth century: is it distinct from elsewhere in Europe?, chapter 3 of this volume.

Clegg, S., Clarke, T. and Ibarra, E. (2001) Millennium management, changing paradigms and organisation studies, *Human Relations* 54(1): 31–6.

Coffey, M. and Langford, D. (1998) The propensity for employee participation by electrical and mechanical trades in the construction industry, *Construction Management and Economics*, 16, 543–52.

ConstructionSkills (2006) *Skills for Productivity 2003–2006*, Bircham Newton.

Crichton, C. (1966) *Interdependence and Uncertainty: A Study of the building industry*, London, Tavistock.

Cully, M., Woodland, S., O'Reilly, A. and Dix, G. (1999) *Britain at Work*, London: Routledge.

Dainty, A. R. J., Bagilhole, B. M. and Neale, R. H. (2002) *Coping with construction culture: A longitudinal case study of a woman's experiences of working on a British construction site,* Fellows, R. and Seymour, D. E. (eds) *Perspectives on Culture in Construction,* CIB Report 275, CIB, Netherlands, 221–37.

Dainty, A., Green, S. and Bagilhole, B. (2007) People and culture in construction: contexts and challenges, Dainty, A., Green, S. and Bagilhole, B. (eds), *People and Culture in Construction*, Spon, London, 3–25.

Davis, C. (2004) *Making Companies Safe: What Works?* London: Centre for Corporate Accountability.

DETR (1998) *Rethinking Construction*. London, Department and Transport and the Regions.

DfEE (2000) *An Assessment of Skill Needs in Construction and Related Industries*, London: Department for Education and Employment.

DiMaggio, P. J. and Powell, W. W. (1983) The iron cage revisited: Institutional isomorphism and collective rationality in organisational fields, *American Sociological Review*, 48, 147–60.

Donaghy, R. (2009) *One Death is Too Many*, Report to the Secretary of State for Work and Pensions. Cmd 7828, London: TSO.

Druker, J., White, G., Hegewisch, A. and Mayne, L. (1996) Between hard and soft HRM: human resource management in the construction industry, *Construction Management and Economics,* 14, 405–16.

Duncan, R., Neale, R. and Bagilhole, B. (2002) Equality of opportunity, family friendliness and UK construction industry culture, in Fellows, R. and Seymour, D. E. (eds) *Perspectives on Culture in Construction,* CIB Report 275, CIB, Netherlands, 238–57.

Emmerson, Sir Harold (1962) *Survey of Problems Before the Construction Industries*, London: HMSO.

EOC (2005) *Free to Choose: Tackling Gender Barriers to Better Jobs* (GB summary report) London: Equal Opportunities Commission.

Felstead, A., Gaillie, D., Green, F. and Zhou, Y. (2007) *Skills at Work 1986–2006*, ESRC Centre for Skills, Knowledge and Organisational Performance (SKOPE), Universities of Oxford and Cardiff.

Fielden, S. L., Davidson, M. J., Gale, A. W. and Davey, C. L. (2000) Women in construction: the untapped resource, *Construction Management and Economics*, 18(1), 113–21.

Fitzgerald, I. (2006) *Organising Migrant Workers in Construction: Experience from the North East of England*, London: TUC.

Forde, C. and MacKenzie, R. (2007) Concrete solutions? Recruitment difficulties and casualisation in the UK construction industry, in Dainty, A., Green, S. and Bagilhole, B. (eds), *People and Culture in Construction: A Reader*, London: Taylor & Francis, 26–38.

Gale, A. W. and M. J. Davidson, Eds. (2006) *Managing Diversity and Equality in Construction* London: Taylor & Francis.

Gann, D. and Senker, P. (1998) Construction skills training for the next millennium, *Construction Management and Economics,* 16(5), 569–80.

Gardner, T., Moynihan, L., Park, H., Wright, P. (2000) Unlocking the black box: Examining the processes through which human resource practices impact business performance, paper presented at the 2000 Academy of Management Meeting, Toronto, ON, Canada.

du Gay, P. and Salaman, G. (1992) *The cult[ure] of the customer, Journal of Management Studies,* 29(5), 615–33.

Greed, C. (2000) Women in the construction professions: achieving critical mass, *Gender, Work and Organization,* 7(3):181

Green, S. D. (2006) The management of projects in the construction industry: context, discourse and self-identity, in *Making Projects Critical* (eds. D. Hodgson and S. Cicmil), Palgrave Macmillan, 232–51.

Green, S. D. and S. May (2005) Lean construction: arenas of enactment, models of diffusion and the meaning of leanness, *Building Research & Information,* 33(6), 498–511.

Green, S. D., Newcombe, R., Fernie, S. and Weller, S. (2004) Learning across business sectors: aspects of human resource management in aerospace and construction, in *Project Procurement for Infrastructure Construction* (eds S. N. Kalidindi and K. Varghese), Narosa, New Delhi, 218–26.

Guest, D. (1987) Human resource management and industrial relations, *Journal of Management Studies,* 24(5), 503–21.

Guest, D. E. (1991) Personnel management: the end of orthodoxy?, *British Journal of Industrial Relations,* 29(2), 149–76.

Guest, D. E. (1999) Human resource management: the workers' verdict, *Human Resource Management Journal,* 9(3), 5–25.

Guest, D. E. (1998) Human resource management, trade unions, and industrial relations, Mabey, C., Salaman, G. and Storey, J. (eds), *Strategic Human Resources Management: A Reader,* Sage, London, 237–50.

Guest, D. E., Miltchie, J., Conway, N. and Sheehan, M. (2003) Human resource management and corporate performance in the UK, *British Journal of Industrial Relations,* 41(2), 291–314.

Hardy, C., Palmer, I. and Phillip, N. (2000) Discourse as a strategic resource, *Human Relations,* 53(9), 1227–48.

Harvey, M. (2001) *Undermining Construction,* London: Institute of Employment Rights.

Harvey, M. (2003) *Privatization, fragmentation and inflexible flexibilization in the UK construction industry,* Building Chaos: An international comparison of deregulation in the construction industry (eds G. Bosch and P. Philips), London: Routledge, 188–209.

Higgin, J. and Jessop, N. (1965) *Communications in the Building Industry,* London: Tavistock.

HSE (2001) *Health and Safety Statistics 2000/1,* HSE Books, Sudbury, Suffolk.

HSE (2003) *Causal Factors in Construction Accidents,* Loughborough University and UMIST, HSE Research Report 156.

Ive, G. (2003) The public client and the construction industries, in *Construction Reports 1944–98,* (eds M. Murray and D. Langford), Oxford: Blackwell, 105–13.

Jones, J. R., Huxtable, C. S., Hodgson, J. T. and Price, M. J. (2003) *Self-Reported Work-Related Illness in 2001/02: Results from a household survey*, HSE, HMSO, Norwich.

Keenoy, T. (1990) HRM: a case of the wolf in sheep's clothing, *Personnel Review*, 19(2), 3–9.

Keep, E. and Rainbird, H. (2000) Towards the learning organisation?, *Personnel Management* (3rd edn), S. Bach and Sisson K. (eds), Oxford, Blackwell, 173–94.

Legge, K. (1978) *Power, Innovation and problem solving, Personnel Management*, London: McGraw Hill.

Legge, K. (1994) Managing culture: fact or fiction, in K. Sisson (ed.), *Personnel Management: A comprehensive guide to theory and practice in Britain*, Oxford: Backwell, 397–433.

Legge, K. (2000) Personnel management in the 'lean organisation', S. Bach and Sisson K. (eds), *Personnel Management: A comprehensive guide to theory and practice*, Blackwell, Oxford.

Legge, K. (2005) *Human Resource Management: Rhetorics and realities*, 10th anniversary edn, Basingstoke: Palgrave Macmillan.

LEK (2009) *Construction in the UK Economy: the Benefits of Investment*, L. E. K. Consulting, London.

Loosemore, M., Dainty, A. and Lingard, H. (2003) *Human Resource Management in Construction Projects: Strategic and operational approaches*, Spon, London.

Marchington, M. and Parker, P. (1990) *Changing Patterns of Employee Relations*, Hemel Hempstead: Harvester Wheatsheaf.

Ness, K. (2010a) 'I know Brendan; he's a good lad': the evaluation of skill in the recruitment and selection of construction workers, in: Egbu, C. (ed.), *Proceedings of the 26th Annual ARCOM Conference*, Leeds, Association of Researchers in Construction Management, Vol. 1, 543–52.

Ness, K. (2010b) The discourse of 'Respect for People' in UK construction, *Construction Management and Economics*, 28(5), 481–93.

Noon, M. (2007) The fatal flaws of diversity and the business case for ethnic minorities, *Work Employment & Society*, 21(4), 773–84.

ODPM (2004) *Skills for Sustainable Communities*, London: Office of the Deputy Prime Minister.

ONS (2010) *Construction Statistics Annuals 2000–2010*, Office of National Statistics, London.

Parker, M. (2002) *Against Management*. Cambridge, Polity Press.

Pepper, C., Dainty, A. R. J., Bagilhole, B. M. and Gibb, A. G. F. (2002) Diversity: a driver of business improvement within the UK construction sector?, Greenwood, D (ed.), *18th Annual ARCOM Conference*, 2–4 September 2002, University of Northumbria, Association of Researchers in Construction Management, Vol. 1, 277–85.

Powell, W. W. and DiMaggio, P. (eds.) (1991) *The New Institutionalism in Organizational Analysis*, University of Chicago Press, Chicago.

Raidén, A., Pye, M. and Cullinane, J. (2007) *The nature of the employment relationship in the UK construction industry*. Dainty, A., Green, S, and Bagilhole, B. (eds), *People and Culture in Construction*, London, Taylor & Francis, 39–55.

Rees, C., Scarbrough, H. and Terry, M. (1996) The People Implications of Leaner Ways of Working, Report by IRRU, University of Warwick, *Issues in People Management, No. 15*, Institute of Personal Development, London, 64–115.

Respect for People Working Group (2000) *A Commitment to People 'Our Biggest Asset', London:* Rethinking Construction.

Respect for People Working Group (2004) *Respect for People: A Framework for Action,* London: Constructing Excellence.

Respect for People Working Group (2007) *Respect for People: The business benefits,* London: Constructing Excellence.

Rousseau, D. (1995) *Psychological Contracts in Organizations,* Thousand Oaks, CA: Sage.

Stinchcombe, A. L. (1959) Bureaucratic and craft administration of production *Administrative Science Quarterly,* 4(2): 168.

Storey, J. and Sisson, K. (1989) The limits to transformation: human resource management in the British Context, *Industrial Relations Journal,* 21 Spring, 60–65.

Tressell, R. (1965) *The Ragged Trousered Philanthropists,* London: Panther.

Truss, C., Gratton, L., Hope-Hailey, V., McGovern, P. and Stiles, P. (1997) Soft and hard models of human resource management: a reappraisal, *Journal of Management Studies,* 34(1), 53–73.

Warhurst, C., Eikhof, D. R. and Haunschild, A. (2008) *Work Less, Live More? Critical Analysis of the Work–Life Boundary,* Basingstoke, Palgrave MacMillan.

Watson, T. J. (1977) *The Personnel Managers: A Study in the Sociology of Work and Employment,* London: Routledge.

Watson, T. J. (1986) *Management, organization and employment strategy,* London: Routledge & Kegan Paul.

Watson, T. J. (1995) Rhetoric, discourse and argument in organisational sense making: a reflexive tale, *Organizational Studies* 16: 805–21.

Watson, T. J. (2004) HRM and critical social science analysis, Journal of Management Studies, 41:3.

Watson, T. J. (2008) Managing Identity: Identity work, personal predicaments and structural circumstances, *Organization,* 15(1), 121–43.

Willmott, H. (1993) Strength is ignorance: Slavery is freedom: Managing culture in modern organisations, *Journal of Management Studies,* 30(4), 515–52.

Winch, G. (1998) The growth of self-employment in British construction, *Construction Management and Economics,* 16, 531–42.

Wood, Sir Kenneth (1975) *The Public Client and the Construction Industries,* London: HMSO.

Wood, S. and de Menezes, L. (1998) High commitment management in the UK: evidence from the workplace industrial relations survey and employers' manpower and skills practices survey, *Human Relations,* 51(4), 485–515.

3 The development of building labour in Britain in the twentieth century: is it distinct from elsewhere in Europe?

Linda Clarke, Charlie McGuire and Christine Wall

Introduction

The purpose of this chapter is to chart the social development of construction labour in Britain, attempting to pinpoint its changing character. It will be seen that it was only after the 1970s that the deregulation of employment conditions in the British construction industry became more and more embedded and that the division of labour, the nature of vocational education and training (VET), and wage and industrial relations became qualitatively different from those of continental countries such as France, Germany and the Netherlands. This chapter explores the reasons for this departure, identifying key moments of change in the formation and conditions of labour in Britain and concludes by suggesting a possible alternative direction which would enable a break with the historical legacy of the British system.

In tracing the social development of construction labour in Britain, distinct historical stages are identifiable, as shown in **Fig. 3.1**. Those characteristics become clearer when compared and contrasted with developments in other leading European countries. This does not mean that the historical process can be cut up into chronological chunks or that there are definite stages of development; history is rather a continuous process of human action, creating its own vehicles and therefore belonging to the world of change and transition (Collingwood 1961). With a long-term view, however, not only is it possible to discern how particular social changes occurred but the possibilities for future action, especially when drawing also on developments elsewhere. Nor does it mean that the characteristics of building labour are nationally exclusive; construction labour is rather mobile and the labour process transcends national borders. The same tendencies are likely to be evident in each country, though to a greater or lesser degree, depending on the social, political, economic and institutional context.

This view of social change is important as without it we might regard the continued trade-based character of construction labour in Britain as somehow impermeable. For what stands out above all in a historical comparison of labour in Britain and, for example Germany, is the different way in which labour is conceptualised, defined generally in relation to a particular range of tasks in the workplace and to work as a specific output of labour, rather than in relation to the capabilities or qualifications of the person (Brockmann *et al.*, 2011). It is a distinction between what Richard Biernacki (1995), in his historical comparison of labour in Britain and Germany, terms 'embodied labour' as distinct from the power of labour itself which he terms 'labour power'. And it is one echoed again in the important 1980s study *Europe et Chantiers*, comparing the construction sectors of Britain, the Federal Republic of Germany (FRG), France and Italy (CEREQ 1991; Campinos-Dubernet and Grando 1991 and 1992). In this study, focussed on types of labour organisation and the means of reproducing skills, productivity in the British construction industry was estimated to be about half that of the FRG and two-thirds that of France (Margirier 1991). Why was this so, why do such differences persist, and how is the character of construction labour today changing?

Labourers and craftsmen before the Second World War

As shown in **Fig. 3.1**, the main issues confronting building labour in Britain in the years up to the Second World War were the length of the working day and piecework. As throughout the nineteenth century, the division of labour continued to rest on the distinction between the craftsman and the labourer, a distinction reflected in union organisation. The apprentice was apprenticed to a trade, though often indistinguishable on site from the labourer. And the key feature separating building labour from labour in many other industries, including manufacturing, was adherence to the time-based wage.

Wage, employment and industrial relations

In the nineteenth century, labour struggles were bound up with attempts to establish a regular working day and standardised payment by the hour, which by the end of the century had become the 'respectable' wage form, predominating in all local working rules, in contrast to piecework (Druker 1991:45). There was a high degree of standardisation of wage rates within and between localities, despite multiple rates throughout the country. Thus, by 1897, whilst the Amalgamated Society of Carpenters and Joiners (ASCJ) had 20 different rates for carpenters, these were fairly uniform irrespective of locality. Standardisation was facilitated in the first place by the 1891 Fair Wages Resolution, that trade union standard rates and conditions should apply on all government contracts and there should be no sub-letting (Bercusson 1978). Then from the very end of the nineteenth century collective bargaining was encouraged by the Conciliation Act 1896, setting up Conciliation Boards to provide a voluntary framework for employers and trade unions to locally regulate wages and conditions of work.

Table 3.1: Historical stages in the development of building labour[1]

	Key issues	Division of labour	Vocational Education & Training (VET)	Wage forms/ relations	Industrial organisation
Free collective bargaining 1890s–1930s	– working day – piecework	craft/ labourer/ apprentice	apprenticeship	time-based	craft and labourer unions, NFBTO and NFBTE
Social partnership? 1940s–1970s	'lump', nationalisation	skilled/ semi-skilled / labourers/ trainees	Government Training Centres (GTCs), Construction Industry Training Board (CITB)	bonus and plus rates, social wage (WRA)	local shop stewards/ national organisation
State regulation 1970s–2000	self/direct-employment, training, CSCS	skill grades, labourer, trainees	day + block release	shift/day rates, decline in social wage	UCATT, TGWU, Amicus
Occupational/ qualified labour market 2000?	entry into labour market, Social protection	different skill levels	college + work-based	individual employment relation	Unite

Before the second world war, though the traditional trades had long been assimilated to capitalism, the division of construction labour in Britain bore all the hallmarks of a craft system, as the skilled worker was still regarded as having abilities and privileges to work with tools related to particular materials and the trade unions continued to be divided along craft lines. By the early twentieth century there were 13 labour federations for the building industry alone and 72 different unions, local and national, each attempting to regulate apprenticeship. In an attempt to rationalise and improve their organisations, in line with the example of the employers (the National Federation of Building Trade Employers (NFBTE)), the various building unions in 1918 set up an umbrella federation, the National Federation of Building Trade Operatives (NFBTO). This body was loosely structured and, whilst it was given the responsibility to negotiate pay and conditions with the employers, any decisions taken were referred back to the individual unions for their consideration. By 1926 negotiation of collective agreements on a national scale was through the National Joint Council for the Building Industry (NJCBI), consisting of employers' and workers' representatives.

The construction industry remained, however, divided into two sectors, building and civil engineering, and by 1919 employers engaged on works such as road, sewerage, water supply, dams, and hydro-electric schemes were arguing that the nature of this work was different and that its needs were not met by the machinery of the building industry. The inability of some of the labourers unions to gain quick acceptance into the NFBTO led one in particular, the Navvies Union, to seek direct negotiations with the employers' body, the Federation of Civil Engineering Contractors (FCEC). This stance was supported by other general unions, leading to the establishment of the Civil Engineering Construction Conciliation Board (CECCB) working rule agreements (Hilton 1968: 137–38).

Collective agreements for the industry covered craftsman, labourer and apprentice rates, hours of work, extra payments, overtime, night gangs, travel (walking and lodging) allowances and other local practices such as the provision of site facilities and tools (Ministry of Works 1950). Minimum weekly working hours stood at 44 in 1920 and were not reduced (to 42) until 1961. Regional up-grading meant that, from 17 skill grades for different areas in 1922, there were only six by 1947. Pay rates were also from 1923 on a sliding scale agreement, adjusted according to the Retail Price Index. During the war the industry was regulated and co-ordinated through a complex committee structure including representation by employer and employee organisations—in effect a period of social partnership. Until this time collective agreements for the construction industry were on the basis of time, not incentive-based. This reflected in particular the craft unions' fundamental opposition to piece-working, a practice seen as synonymous with 'scamped', poor quality work, and the deterioration of their craft, as well as the setting of workers against each other (Wood 1979: 63). Indeed, as summarised in **Table 3.1**, until the

Second World War the building industry represented a 'great example of a time-work industry', with a uniform rate for the whole country and a fixed differential between craftsmen and labourers (Cole 1918: 1113).

Occupational divisions, apprenticeship and training

The employment relation rested essentially on a free contract between the employer and the workforce, including the apprentices. This implied a lack of obligation on the part of employers with regard to apprenticeship which was, as a result, as in the nineteenth century, bound up with the development and strength of the trade unions and collective bargaining (Howell 1877). With the formation of the National Wages and Conditions Council for the Building Industry in 1921, the institutional basis was set for National Apprenticeship Agreements aimed at the joint regulation of industrial training at national and local levels through collectively agreed rates. And by the 1930s and 1940s the trade unions were proposing joint regulation, the attachment of the apprentice to industry rather than individual employers, the involvement of education authorities, and day release as a right. The National Standard Apprenticeship Scheme, agreed during the Second World War by the NFBTO and the NFBTE, represented the culmination of the collectively agreed system. The scheme was mainly for bricklayers and carpenters, extended for four to five years, and involved day release, no overtime, national and local regulation through the National Joint Apprenticeship Board and a trust fund should the apprenticeship break down. But the continued decline in apprenticeships on into the post-World War II period demonstrated the inadequacy of a voluntary system, in spite of this more coordinated effort (Clarke 1999).

In November 1941 G. D. H. Cole presented a wide-ranging report highly critical of the existing state of training.[2] He recommended:

> a form of apprenticeship to the trade as a whole rather than to an individual employer may have much to recommend it.... and that any satisfactory scheme must include a much larger element of formal technical instruction in schools or classes than is usual in practice at the present time.[2]

He also urged the recognition of a wider range of occupations as involving 'special types of skill'. One key problem was that there was no clear means for the traditional craft occupations to adapt to new construction activities associated with, for example, concreting. The only method was for new processes to be 'claimed' by a craft union as within their potential remit, such as claims by plasterers and bricklayers that cement floors and breeze-block partitions were within their trade (Hilton 1963; Postgate 1923). The Simon Committee report of 1943 noted the importance of scientific education to enable the skilled worker to adapt to new working conditions, and of a broad general education. It also foresaw a 'trend towards public responsibility for apprenticeship as an institution'.

However, such radical proposals did not appear in the White Paper (1943), *Training for the Building Industry* which proposed a construction workforce of 1.25 million, necessary to undertake a programme of reconstruction stretching over ten to twelve years, an end to casual employment and the training of 200,000 men on intensive government courses funded by the state to alleviate shortages in the years immediately following the war. The Act defined the terms that were to continue to distance apprenticeship from mainstream education:

> Apprenticeship training, unlike special adult training, will not be provided and paid for by the State and the various questions which arise in controlling apprenticeship are traditionally settled by the industry itself (ibid., 13).

The White Paper endorsed apprenticeship training as 'the recognised method of training in employment and of entry into the ranks of the skilled workers' and established the Building Apprenticeship and Training Council (BATC), consisting of employee and industry representatives together with members of various government departments. Educationalists played a minimal role: out of a total of 53 members of the BATC, only five represented educational bodies. No attention was given to the widening of training to take in new and emerging construction occupations or to the problem of having very high numbers of labourers in the industry. Studies were however undertaken to estimate the numbers needed to be recruited into the skilled trades, with agreement on a ratio of one apprentice to every four skilled workers in 'normal times' – training levels not ever achieved in the 1950s and 1960s (**Table 3.1**; BATC 1944). The ensuing 1944 Education Act contained provisions for local authorities to provide further training and for the setting up of National Joint Apprenticeship Boards so that apprentices were educated in the employer's time through day release once a week to a technical college.

Parallels with France and Germany

Up until the second world war advances in the development of building labour in Britain appear, if anything, advanced compared with other European countries, with unions playing an important role in determining a 'norm' for the working day, a uniform wage rate and relatively progressive wage relations. In France, the industry was dogged in the 1890s by the extensive use of *tâcheronat*, whereby the employer paid for a definite task, contractually defined, to a *tâcheron* who in turn hired workers, a system comparable to labour-only subcontracting and which introduced similar problems with respect to sacrificing quality because of the need for intensity of work and frequent failure to receive payment (Ribeill 1991). The 1899 Millerand Decree was, however, equivalent to the Fair Wages Resolution of 1891 in demanding an equal wage rate for public works in any one locality or region (Tallard 1991a). Similar to Britain, too, collective agreements were introduced in France between 1919 and 1932, legalising rates according to trade and locality, though the rates were very numerous and only applied nationally from 1926.

Parallels to Britain are also evident in Germany where, by the end of the nineteenth century, the building unions were top of the league in terms of wage rates, reductions in working time and the development of collective agreements. New wage systems were introduced associated with new areas of work and, as early as 1890, accident and sickness insurance and pension schemes existed (Schudlich 1991; Weis 1991). Apprenticeship training in Germany had initially been the prerogative of the craft guilds, who administered provision and certification, but the system was transformed in the early twentieth century and the model for the post-War Dual System was based instead on the system of industrial training established by highly capitalised firms such as Siemens (Hanf 2007). Dramatic advances were made following the 1918 Revolution, with universal social welfare for the unemployed, a collective labour law recognising collective agreements on working conditions, time-rate wages for all trades and the introduction of an eight-hour day. In 1919 priorities set by the German Association of Construction Workers included: the eradication of payment by results (PBR) in the form of *Akkordarbeit,* the erosion of wage differentials, and complete unionisation (Weis 1991). As in Britain too, there was a national building strike in 1924, resulting in wage increases. From then on, however, parallels with Britain cease as regional and skilled-labourer differentials increased, work was made a condition of claiming unemployment benefit and the vast majority of construction union members were classified as unemployed.

A collective post-War building labour force?

With the end of the Second World War, many of the characteristics of building labour in Britain in the pre-War period disappeared, as shown in **Table 3.1**. Instead of the working day and piecework, the key issues became 'lump labour' (see below) and even the nationalisation of the building industry, as the bonus and 'plus rates' took a hold of the time-based wage structure, threatening to dismantle the careful construction of a social wage and bringing local shop stewards and the national unions into conflict. The division of labour in the industry too was transformed as a range of new occupations classified as 'semi-skilled' were introduced, accompanying technological changes, intruding on the traditional craft/labour dichotomy. The government, forced to respond with new schemes and regulation of training, struggled to keep pace with developments.

The changing nature of building activity

At the beginning of the post-Second World War period, there were an estimated 130,000 building firms in Britain, reducing to around 73,000 by the end of 1960s, accompanying the rising number of large firms. In 1968, 26 per cent of the workforce worked for firms that employed over 500 people; at the same time 78 per cent of all building firms – around 66,000 – employed between one and ten operatives, and 50 per cent of workers were employed by firms of under 100 operatives (Hilton 1968: 10). On many of the contracts held by the major firms direct employment was low and sub-contracting rife.

Changes in the size of firms went hand-in-hand with significant changes in the nature of building activity, with much more widespread use of new production methods during the war. In the post-war period, 'non-traditional' techniques of building, designed to be executed with fewer numbers of skilled workers, along with more sophisticated and better machinery and technology became more common. In 1951, the first tower crane was used in Britain and mechanical excavators and diggers became increasingly large and more powerful. Concrete specialisms, such as steelfixing and concrete-finishing also became more evident on sites. In 1954, around 20 per cent of all new local authority funded houses in Britain were systems built (Bowley 1966: 202).[4] Traditional craftsmen were not rendered obsolete and did adapt to the new conditions, but the nature of their work changed and was not always classed as 'skilled' (Austrin 1978).

Another development in the post-War period was the emergence of extremely large industrial building projects. Large oil, coal and nuclear power stations, oil refinery and chemical works projects from 1950 onwards tended to blur the divide between civil engineering and building. Large buildings associated with the emerging welfare state, huge motorway projects and a substantial growth in house-building—which rose steadily from around 140,000 units per year in 1947 to around 500,000 by 1970 – all signified the changing nature of building activity.

The development of a social wage

Along with changes in the nature of firms in the industry in the post-War years went a series of radical reforms in wage and employment relations. By 1945, a uniform rate of pay for the whole country and a fixed differential between craftsmen and labourers of 80 per cent[5] were significant aspects of the building industry in Britain (NBPI 1968). Neither the ASW nor the Amalgamated Union of Building Trade Workers (AUBTW) accepted the contention that apparently low levels of productivity in the building industry were due to time-based contracts but rather argued that PBR represented a 'degrading process of being sized up, like a racehorse' (New Builders Leader 1941; Hilton 1963: 277). Indeed, rather than introduce bonus systems, the unions considered that the answer to the problems of the building industry lay in nationalisation (Connelly 1960: 100). Despite this, the post-war government, pressed by the estimated need to build 1.25 million homes, intervened on the side of the employers and suggested a new incentive system that would pay construction workers 20 per cent above the basic rate and give a total increase of 3d per hour (Allen 1952: 259). Thus, in 1947 PBR was accepted generally by the NFBTO in return for a wage increase. In his history of the ASW, Connelly argues that the 1947 agreement was different from piecework because operatives were still guaranteed a basic rate of pay and that 'where a standard rate exists, the super-imposing of bonus schemes offers no fundamental problems, and hardly any problems at all where unions are strong enough at the point where the schemes are negotiated' (Connelly 1960: 103). The basic rate of pay remained low, however, and the length of the working week much the same as before the war; it was not until 1966 that the 40-hour week was achieved. The bonus became an ever-increasing proportion of the wage (up to 100%), wage differentials widened and by the 1960s a serious wage drift had developed (Allen 1952; Handy 1971).

Though the acceptance of PBR was to have a devastating impact on the wage structure, the agreement on annual paid holidays, administered through a process of holiday stamps,[6] contributed to what was to become a striking feature of the post-War situation – the development of the 'social wage', which meant the increased importance of indirect payments in the wage package. Holiday pay was premised on a guaranteed working week of 30 hours, increased to 32 by 1947, which meant the establishment of a regular system of employment in the industry throughout the year and that employees could no longer be dismissed because of bad weather (by 1948) – because of the breakdown of plant, non-arrival of materials, etc. Prior to this, despite demands by the trade unions, there had been no pay for such stoppages (Allen 1952). Additional wage components, constituting 10 per cent of the wage cost and including also tool allowances, were coupled with the post-war extension of the national insurance system and with the introduction of industry-based pension and sick pay schemes, restricted though to key personnel and not transferable between employers (Girdwood Report 1948; UMIST 1973).

The effect on earnings of the acceptance of PBR and the development of the social wage was mixed. Whilst in 1938, average earnings for workers under NJCBI agreements were six per cent more per hour than for those under civil engineering agreements, by 1959 this situation had reversed and civil engineering workers were earning six per cent more per hour (than those under the NJCBI agreement; Hilton 1968: 144). The working rules of the CECCB were characterised by plus rates, which were additional payments to men carrying out work that required particular 'skills', responsibility or risk, including crane operation, digger, mechanical shovel or dumper driving, concrete mixer operation, etc. Civil engineering continued to be based on the non-skilled 'labourer' rate (i.e not awarded 'craft' status); nor was training formalised in any structured way. This new, changing situation caused huge problems on sites, notably on the 1951 Festival of Britain site, where non-unionised labourers were earning more than unionised workers operating under the NJCBI agreements. By 1967 the civil engineering agreement had around 140 plus rates, 23 of which were high enough to give those workers a higher wage than craft workers.[7]

Industrial relations in the industry

The 1950s and 1960s were decades of great change in industrial relations in the industry, marked by the gradual erosion of the fragile social partnership seen during the war. Two clear rifts began to emerge: between the employers and employees and between the national unions and rank-and-file members. These were exacerbated both by the introduction of PBR and widespread use of labour-only sub-contracting (LOSC) or 'lump labour'.

Lump labour

The lump was a form of wage contract where the contractor hired on a labour-only basis and paid workers an agreed lump sum for an agreed amount of work. It was this system above all which undermined regulated wages and meant that the full 'social wage' ceased to progress. Though 'self-employed' workers performed the same tasks on site as directly employed workers, they had a different legal and tax status, which had severe implications for trade union organisation. The contractor hiring the lump labour had no responsibilities, in terms of income tax deduction, the payment of National Insurance contributions, holiday pay, sick pay or any of the other costs normally paid by an employer – representing what amounted to an employment subsidy. The decision in 1971 to legitimise the lump through the issue of special tax certificates, numbering over 350,000 by October 1972 and allowing operatives to be responsible for their own tax and insurance or to have these deducted at source, only facilitated the conversion from direct to self-employment and represented a further inducement to firms to work outside the recognised framework (Wood 1979). Those covered by these certificates came under a contract for services outside the social wage and without employment protection. As a result, whilst the basic rate of pay for directly employed workers stipulated in the 1971 NJCBI agreement was £20 per week for craftsmen and £17 for labourers, the wage paid to those on the lump could be much higher though labour costs were lower. Lump workers also had an incentive to get the work finished as quickly as possible, often had little interest in health and safety considerations, site conditions or the training of apprentices, and little reason to join a union.

The negative effects of the lump on trade union organisation soon became evident. The post-war decades of labour shortages should have been years when the strength of the building unions increased and union membership rose. Instead, from the 1950s onwards craft union membership declined so that by 1965 only 28 per cent of the total building labour force was in NFBTO-affiliated unions (Hilton 1968: 72). The number of unionised carpenters and joiners dropped from 198,365 in 1950, to 189,233 in 1959 and was just 125,457 by 1974, despite this being a time of growth in the building industry with an increase in their overall numbers. The situation was even worse for bricklayers, despite a temporary influx of new members following amalgamation with the National Building Labourers in 1952 (Hilton 1963). In 1947, there were 90,550 members of the AUBTW, but by 1964 the number of unionised bricklayers had dropped to 59,224 and, ten years later, the figure was at only 44,120 (Austrin 1978: 176). Meanwhile another 'wet-trades' union, the plasterers, was also suffering, with its membership declining from 17,762 in 1948 to 12,567 in 1964 (Wood 1979: 84). In belated recognition of the considerable problems faced, the three main craft unions, the AUBTW, ASW and Amalgamated Society of Painters and Decorators (ASPD) finally amalgamated in 1971 to form the Union of Construction and Allied Technical Trades (UCATT).

PBR also impacted on industrial relations. One effect of the introduction of incentive payments was to place far more power in the hands of site stewards who would often negotiate bonus arrangements. In the 1960s, this became a feature of industrial relations, particularly on large contracts, and shop stewards committees, comprising representatives of craft and non-craft unions alike, took the place of the official union negotiators. On some sites where union organisation was strong, collective site bonus agreements were reached, with all workers – regardless of personal performance – receiving an equal share, though this was resisted by employers, who preferred individual bonus arrangements. Greater control over the hiring and firing of labour was also sought and an end to the use of 'lump' labour. These tendencies were evident in, for example, the building of Stevenage New Town, where strong union organisation at site level ensured that between the 1950s and the 1970s no lump workers were allowed onto development corporation contracts.[8] The increasingly casual and unregulated nature of much employment in the industry by the 1960s was underlined in the findings of a survey, that 43 per cent of employees of main contractors had less than six months with the firm and were, therefore, subject to one day's notice (Jeanes 1966) In 1965 the lump was estimated to include 200,000 operatives and by 1973 those classified as 'self-employed' numbered more than 400,000 out of a total workforce of about 1.2 million, mainly in traditional trades such as bricklaying, floor-laying, roofing, plastering, carpentry and joinery and painting and decorating (Austrin 1980; Wood 1979).

As Austrin (1978) points out, the failure of the craft unions to halt the lump and the hugely destructive effect this form of wage contract had on building unionism created the conditions for the unofficial, rank-and-file shop stewards movement, which was to be critical to the future of building trade unionism and came into direct confrontation with the union leadership as in the disputes over bonus payments on the Barbican site in the mid-1960s. The Barbican strike was in many ways a hollow victory for the union leadership at national level in that it exposed the chasm that existed between it and the membership. In the period following the strike, the Charter movement developed and established a network of rank-and-file bodies in several major towns and cities in Britain. Its main demands were: a wage of £35 per week, a working week of 35 hours, an end to lump labour, the de-casualisation of the industry, protection of shop stewards from victimisation, a single industrial trade union, and improvements to health and safety on sites. Reflecting the pressure placed on it by the rank-and-file organisations, UCATT lodged this wage claim, which was rejected by the employers and a strike at selected sites began in June 1972 (Building Workers Charter 1972). A compromise deal was eventually reached for a weekly wage of £26 (but with no reduction in working hours), the biggest single increase the building unions had ever achieved.[9] In the period following the strike, Charter supporters entered the UCATT executive and their conference delegates were able to overturn official union policies on wages, union structure and position on the Shrewsbury Pickets, who had been convicted and imprisoned in relation to events that had occurred during the strike.

The development of qualified labour

In terms of the formation of labour, the key features of the post-War system were: the distancing of the state from VET; concentration on apprenticeship to the traditional trades; reluctance to establish a comprehensive scheme of training outside these occupations; and the constant reaffirmation of the division between education and practical knowledge, between knowledge acquired outside the workplace and within (Cotgrove 1958: 33). The BATC met regularly and attempted to systematise and standardise apprenticeship training through the setting up of the National Joint Apprentice Scheme for the building industry in 1946 to cover the main crafts, operated throughout England and Wales by Joint Apprentice Committees consisting of employers, union, education, and government representatives. The registering of apprentices under a standard form of indenture and the issuing of certificates of completion were promoted and administered by the NJCBI on a voluntary basis. This met with only a certain degree of success: in 1951 only 60 per cent of industry apprentices were registered, declining by 1954 to 53 per cent; the rest remained under verbal agreements with their employer. Post-war reconstruction plans, to train 260,000 building trade operatives in six-month schemes in Government Training Centres (GTCs), were initially resisted by the unions, but eventually accepted, on the proviso that they were under the control of the newly-established national apprenticeship scheme committees.

The lack of a statutory basis made the whole system difficult to implement and BATC was finally disbanded in 1956 claiming to have established a 'firm foundation' for apprenticeship training in the industry but acknowledging that problems 'remain many and varied.'[10] The 1956 White Paper on Technical Education did outline plans for a massive injection of funds into further education and for major reorganisation. However, until the Construction Industry Training Board (CITB) was established in 1964, training was again left in the hands of the industry where 'custom and practice' masked the ingrained prejudices of both trade unionists and employers (Clarke and Wall 2011). Though this was one of the most intense periods of technological and occupational change, it was not until 1963 that the NJCBI commissioned the BRS to run a three-year investigation into building operatives' work on traditional and non-traditional sites throughout England, Scotland and Wales. A total of 5,343 operatives were interviewed, including foremen, electricians, gangers, clerks of works and others and the results were published in 1966 in two volumes, *Building Operatives Work Volumes 1 and 2* (Jeanes 1966). The report presented in great detail a trade-by-trade review of the actual work that was being carried out on building sites, both traditional and systems build. One of the main findings, that the work apprentices experienced on site was entirely due to chance and not the result of any planned scheme of site training, was a clear indictment of the unregulated and unstructured training available in the industry. Another was that, although 75 per cent of carpenters and joiners, bricklayers and plumbers had attended college as part of their apprenticeship, few had gained any qualification. Only 43 per cent of carpenters, 39 per cent of bricklayers and 50 per cent of plumbers held a craft certificate. Finally, nearly half the workforce was found to be composed of non-apprenticeship trades or occupations and thus not in receipt of any formal training whatsoever. The report not only revealed the extent of the training crisis in the industry but also the range of occupations found on sites, numbering over 40, despite industry statistics still only recognising seven building trades: carpenters and joiners, bricklayers, slaters and tilers, plasterers, painters, plumbers and glaziers, and masons (not including electricians and heating and ventilating engineers).

It was this situation which confronted the newly-established CITB, set up under the Industrial Training Act of 1964 and representing 'the first attempt to formulate a modern industrial manpower policy', (Perry 1976: xix). The Act sought to give trade unions a fuller role in training policy through the establishment of tripartite statutory Industrial Training Boards (ITBs), numbering 27 by 1969. This imposed an obligation on employers to train through the institution of a levy-grant system, whereby all firms except those below a certain size paid a training levy to the CITB and those providing training received grants. The majority of training in construction was undertaken by small and medium-sized building firms (**Table 3.1**). In 1957, 75 per cent of all trainees were with smaller firms and this proportion was still high, at 70 per cent, in 1965 though declining to 63 per cent in 1970. The remit of the ITBs was to: establish policy with respect to training, including its length, the registration of trainees and their attendance at a further education college; set standards of training and syllabuses; provide advice and assistance about training; devise tests to be taken by apprentices and instructors; run training course in training centres; pay grants to reimburse firms and allowances to trainees not taken on by firms; collect levies; and borrow. Nevertheless, they had a number of important drawbacks. In the first place, they were not closely connected with the education system. Employers too were wary about the degree of coercion that they represented. And the trade unions were not consistently in favour either. Indeed the craft unions were generally unenthusiastic as they continued to favour entry through traditional apprenticeship since that route acted as an important control on labour market entry and on the maintenance of pay differentials. The general unions, however, seeing opportunities for their members where none had existed before, were more enthusiastic.

Despite the findings of the BRS research, the CITB did not immediately attempt to overhaul training radically even though it acknowledged that this was demanded by 'the accelerating pace of technical progress' (CITB 1966: 7). Instead it developed onsite training schedules for apprentices, gave employers guidelines on standards and their responsibilities towards trainees, and reiterated the habitual mantra of the industry minimising the importance of educational input:

> Traditionally, training for the construction trades has always been on the job, the master or competent operator training the newcomer, whether adult or apprentice. The introduction of widespread day-release studies and training for apprentices during the last 20 years has been a major achievement of which the industry is justly proud ... This is, however, a supplement to training on the job, not a replacement of it. It is widely held that no replacement is possible and against this background the programme of training development has to be planned (CITB 1967:13).

The CITB did, nevertheless, initiate a long-term programme to review the occupational structure and training needs and in 1969 introduced the New Pattern of Operative Training (NPOT), a plan of training devised to encompass all areas of building activity. This programme finally attended to the content of training rather than only the length, with the CITB producing all the required training material for use in Colleges and other recognised centres and on site (CITB 1969). NPOT was structured around a first phase spent in basic training, after a short introduction common to all construction occupations, consisting of alternating blocks of time spent in a training centre or college and on site. The second phase was composed of a selection of modules in a single trade, again comprising on- and off-the-job training bringing trainees to a level of competence acceptable to the industry. A third, optional phase consisted of advanced modules in a chosen trade that could be taken at any time over the working life of an operative, and used as a way of keeping abreast of technical change. A separate training plan for the general construction operative, aimed at semi- and unskilled workers but of a shorter duration, was also introduced, comprising a first phase of basic training followed by modules in a range of activities such as scaffolding, steel fixing and bending, plant operations, formwork carpentry etc.

Perhaps the most revolutionary aspect of NPOT was the new grouping of occupations with a common core curriculum in the first year so that trainees in, for example, the wood trades followed the same introductory course, with specialisation in carpentry and joinery, wood machining, or formwork carpentry, occurring only in the second year. However, the scheme lasted only a short time because of opposition from both employers and unions and, together with the full-time integrated course, was phased out (Burchell 1980). Under the Standard Scheme of Training set up by the CITB in the 1970s and approved by the NJCBI, formal block-release vocational training was integrated into work-based training, allowing for the gradual reduction of training as a skilled worker to three years. A 'complete package' of operative training was fully introduced in 1974[11] as New Entrant Training with all trainees spending the first six months full-time in college, or in one of the local CITB training centres, before going on site. Although offered high-quality, monitored training, trainees employed by firms paying a training levy to the CITB remained a minority among all construction trainees. The majority of training in the construction industry remained *ad hoc* with either no agreement or only verbal agreements between employer and trainee.

The amount of training began to decline (having first risen with the setting up of the CITB) – particularly in the larger firms where the number of craftsmen per apprentice remained high – at 15 in 1965 in firms with over 1,000 employees compared with four in firms of under ten employees (**Table 3.2**). Training effort remained concentrated on the trades of carpentry and joinery, bricklaying, roofing, plastering, painting, plumbing, heating and ventilation, and electrical work. The number of construction apprentices rapidly declined, from 135,000 in 1965 to 80,000 by 1975 and 66,000 by 1978 – with formally registered apprentices falling even more dramatically, from 37,000 in 1974 to 17,000 in 1984 (Annual Census). At the same time, the number of workers in occupations without a recognised training route or method of entry rose to 53.4 per cent by 1970, as the industry became more reliant on mechanisation and new processes, and the number of those classed as labourers decreased by about 62 per cent between 1965 and 1984, from nearly 400,000 to about 150,000 (Clarke 1992a). In effect there was a dramatic decline in the numbers employed in the traditional trades, with the employment of carpenters and bricklayers, for example, falling by 50 per cent in the 20 years from 1965 and painters by 60 per cent. These were replaced by newer occupations such as concreting, formwork, steelbending,

Table 3.2. Number and proportion of trainees by size of firm 1957–1970

Size of firm (no. of operatives employed)	1957		1960		1963		1966		1970	
	Trainees	Craftsmen per apprentice	Trainees	Craftsmen per apprentice	Trainees	Craftsmen per apprentice	Trainees	Craftsmen per apprentice	Trainees	Craftsmen per apprentice*
1–10	22,835	5.0	20,289	5.1	20,814	5.0	25,366	5.0	15,693	4.1
11–99	37,393	5.9	36,123	5.6	41,709	5.4	60,884	5.4	32,069	4.7
100–499	14,121	8.0	13,871	7.5	16,272	7.0	20,576	7.0	14,288	7.2
500–999	2,473	13.4	2,337	10.0	3,841	10.4	5,892	10.4	4,786	12.3
1,000+	3,327	17.3	3,911	14.0	5,417	12.1	9,009	12.1	7,498	15.0
Total trainees	80,149	6.7	76,531	6.3	88,053	6.2	111,718	6.2	75,234	6.1

Sources: Data for 1957–63 from Collection of construction statistics (1971) BRS; data for 1966–70 from unpublished MPBW statistics; data on craftsmen per apprentice from Table 7.20 Collection of construction statistics c. 1969, BRS, MPBW. 'Trainees' here indicates all those undergoing some form of training including indentured apprentices and those with informal agreements.

[1] The total numbers of trainees between the years 1957 and 1965 from the first source do not tally with unpublished MPBW statistics where totals for these years are higher so that this table cannot be read as a continuous historical series:[2] MPBW recognised 11 categories of firm size, though these have been condensed in this Table for ease of comparison with Ministry of Works statistics. From 1966, however, size of firm categories as used by MPBW change: 1–10 becomes 2–13; 11–99 becomes 14–114; 100–499 becomes 115–599; 500–999 becomes 600–1199; and 1,000 and over becomes 1200 and over.

* figures for 1965

and crane driving, all without a recognised apprenticeship. Though CITB efforts centred on traditional building occupations, by 1978 34 different occupations were listed for construction on their levy-grant form (CITB Annual). A major effect was to deter the social development of labour, cementing old divisions and enforcing trade sectarianism as training became fossilised more and more around these old trade divisions.

Continental parallels and differences

The one major attempt to reform the system of skill formation in Britain, the Industrial Training Act of 1964 preceded the 1969 Vocational Training Act (*Berufsbildungsgesetz*) of the Federal republic of Germany (FRG) by five years. In many respects, the aims of the two Acts were comparable, though it was not long before considerable divergence was apparent in the nature of VET. Just as in the British case, at the heart of the FRG VET system lay the principle of self-government, though this was on an industry rather than a trade basis, with unions and employers' associations organised into 16 industrial sectors established nationally after the War. The dualism of the West German system became established when education reformers were successful in their calls for a compulsory learning element to complement workplace training (Greinert, 2007). It was through the initiative of industry that VET became more formalised and standardised training schemes and assessment criteria were laid down. Importantly, the division of labour in the workplace and the relevant training schemes were reorganised through the principle of *Beruf*, associated with a recognised social status and identity (Streeck and Hilbert 1991). The institutional framework of the FRG VET system also allowed for cooperation between craft (*Handwerk*) and industry bodies, while increasingly powerful industrial unions were granted the right to participate on the boards of various chambers and played a key role in developing training ordinances (Deissinger 2008). The training occupations (*Berufe*) covered a comprehensive and clearly demarcated body of systematically related activities, theoretical knowledge and skills. The qualification, awarded on completion of a standardised programme of training, denoted a level of skills and knowledge enabling occupational mobility.

Observations made by a joint team of employers and trade unions from Germany, who visited Britain on behalf of the OECD in 1964 to examine the VET system, provide key insights into the differences between the FRG and Britain (OECD 1964). The delegation expressed surprise at the lack of state intervention in Britain, at training being left, to all intents and purposes, with employers and unions, and at the adherence to traditional time-served, craft-based apprenticeship:

> Both sides of industry are frequently unable to free themselves of the traditional notion that special skills can only be gained through experience.

> It is often hard to convince them that systematic teaching and learning methods can considerably shorten the time required to instil certain forms of knowledge. In this respect, it will be observed that the termination of apprenticeship in Britain is not determined by the successful passing of an examination but solely on the basis of duration of training (at p. 15)

The conclusion drawn was that management-labour relations and the training systems of FRG and Britain were 'too far apart' (OECD, 1964: 15).

The system established in the FRG contrasted with that in Britain in having considerably more state intervention, with the socially-recognised skills of the different *Berufsgruppen* regulated by statutory institutions, including the vocational schools (*Berufsschule*), training workshops, accredited firms and tripartite examination commissions. The unions and employers were industrial, not divided into general, industry and craft unions each fighting to maintain control over entry into the labour market through apprenticeship, and the construction sector was covered by one large union, IGBau. This negotiated one collective agreement, which in 1973 consisted of five different wage groups each related to a different level of qualification (and hence training). There was, therefore, a clear link between the VET system and the system of collective bargaining and an incentive for individuals to upskill to a higher grade.

A crucial difference between the British and the FRG system was the apprenticeship system. Under the FRG Vocational Training Act, the system of *Lehre* reminiscent of the traditional master-apprentice relationship was substituted for *Ausbildung* in a graded training system. Through an agreement of 1979, the *Stufenausbildung* system was introduced for the construction industry, whereby trainees followed a regulated programme linked to curricula comprising a broad knowledge and skills base (including mathematics and physics) with the first year covering a number of construction occupations and with specialisation only in the second and third years. The resulting scope of activities encompassed in different construction occupations, such as bricklayers, became thereby far broader than in Britain, with a high level of autonomy, preparing the holder to carry out complex tasks in relation to the wider labour and work processes.

In terms of employment and wage relations, too, the gulf between Britain and its continental neighbours became wider from the late 1960s. In FRG and France the further development of social components of the wage (e.g. holidays, sickness benefits, bad weather compensation etc.) meant that direct wages constituted a decreasing part of the total labour cost so that the wage became: 'both the price attributed to a given task and the cost of maintaining a collectively guaranteed social status' (Campinos-Dubernet and Grando 1992: 28). In France, as in Britain, holiday pay in the construction sector was generally established after the War and a series of agreements drawn up intended to stabilise employment, offering benefits including holiday, sickness, accident compensation, seasonal and bad weather payments, pensions, unemployment benefits, further education, monthly pay and grade classifications (Tallard 1991a and b). As a result, by 1978 direct wages in the construction industry constituted only 51 per cent of the total labour costs (Pellegrini 1990: 97). Joint employer–worker commissions, often peculiar to construction, were developed, financed through employers' contributions and responsible for pension and provident funds, training, the prevention of accidents at work, profit-sharing and insurances (Lanove 1990). Even greater divergence occurred with the FRG as an annual arrangement of payment for working hours was introduced and the extent of social charges increased from 37 per cent of total labour costs to 50% by 1990, to include winter and weather compensation, a '13th-month' salary,[12] sickness, pension, holidays, vocational training, travel and special payments (Janssen 1992; Lanove 1990; Pellegrini 1990). In this way, even from the 1950s, developments in the FRG increasingly departed from those in Britain, with the German picture characterised by a minimum use of PBR (monitored by the Works Councils), a high level of social security, the elimination of regional and trade differentials, and the overcoming of the seasonal character of construction.

Individualised employment of labour

From the 1970s, with the further decline in direct employment, collective bargaining and training, any progressive development of building labour in Britain appeared to stall. The state's refusal to take direct responsibility for training, adherence to increasingly inappropriate trade divisions, the growth of the 'lump' and labour-only subcontracting, and a serious drift between the collectively agreed wage and the actual wage paid, all contributed to the collapse by the last third of the twentieth century of the collectively agreed and voluntarist system. As shown in **Table 3.1**, the key issues became, instead of lump labour and nationalisation, the growth of 'bogus' self-employment, the protection of direct employment and with it what remained of the social wage, and the registration of building workers. Though industrially-recognised schemes of VET for the traditional occupations were maintained throughout the 1980s, by the 1990s block release to FE college had generally given way to day release and apprenticeship was in sharp decline. Rather too late, skilled grades came to encompass all building workers though these still did not reflect qualification levels and were increasingly marginal to the shift and day rates which dominated payment on sites. Rivalry between the three main unions – UCATT, TGWU and Amicus – also deterred the collective effort to develop building labour, including through schemes of training.

The development of self-employment and the 'lump'

Direct employment was more and more substituted by labour-only subcontracting or 'lump' labour, whether for groundworks, brickwork, carpentry, finishing, etc., leaving only a few workers directly employed in each firm to fetch and carry for subcontractors or set the pace of work. Numbers of self-employed remained nevertheless relatively stable in the 1970s and it was only in the 1980s that they began to rise dramatically to reach 48 per cent of total employment in the private sector by 1986 and 57 per cent or 718,000 by 1990, levels far higher than other European countries where the self-employed constituted less than 10 per cent of the workforce (Winch 1992). The progression of the lump and self-employment can be regarded as an outcome of the increasing individualisation and importance of the bonus system and of the development of subcontracting. From the late 1970s however, bonus bore only a loose relation to output and had become more and more arbitrary and directly-employed operatives on time rates received, more often than bonus payments, the equivalent of a task rate, with hours bargained on a job-and-finish basis, or a basic with an individual top-up, or simply measured day work (Clarke 1992b).

In stark contrast to other leading EU countries the British construction industry had by 1990 the highest proportion (75%) of direct wages relative to total labour cost, the lowest social charges, the lowest total hourly wage cost, low gross wages in relation to purchasing power and a falling position in the earnings' league for all industrial sectors, extreme fluctuations in earnings, and the longest working hours (Lanove 1990; Pellegrini 1990; EFBWW 1991). Those payments that did continue to be regulated through the – albeit voluntary – collective agreement (such as for bad weather, a guaranteed working week, travel time, indemnities in case of injury or retirement and redundancy, and sickness) were also more restricted than those in other leading countries (Druker 1991). For example, numbers on the holidays-with-pay scheme, paying 21 days annual holidays on the basis of 47/48 working weeks of stamps, declined between 1980 and 1990 by 40 per cent, reflecting the 'diminishing relevance of the national agreement to operatives at site level' (Druker 1992:6). In effect by the 1980s great disparities in wage forms were evident even on one site with simultaneous use of: direct employment at time rates; self-employment at task rates; and casual employment at day rates or shift work – involving employers simply deducting 25 per cent of the wage for tax and insurance. The fundamental division of self-employment from direct employment was one reason why UCATT refrained from recruiting self-employed workers until 1988 and it was only in 1990 that the NJCBI included provision for the self-employed. By this time, however, the wage structure had been effectively deregulated, with extreme variations in wages and sharp fluctuations over time (NEDO 1991).

Trade rather than industry VET

Key characteristics of the unregulated wage form are its relative indifference to skill and wide differentials, given that employment is essentially for a given task, irrespective of the potential range of abilities embodied in the worker carrying it out. To return to the distinction made by Biernacki (1995), it epitomises 'embodied' labour as opposed to labour power. This is reflected in the absence of a comprehensive VET system for construction in Britain and often severe skill shortages. In 1966 at the beginning of the CITB the number of apprentices in the industry was 112,000: by 1985 the number of CITB YT entrants stood at 16,400, a figure reduced by almost half ten years later. This fall directly correlates with the decline in direct employees.

Originally apprentices had to register with the NJCBI because apprenticeship was an integral part of the collective agreement. The Modern Apprenticeship of the 1990s, however, was no longer regulated through collective agreements and therefore outside the control of the trade unions. The training levy system too, based by the early 1990s on 1 per cent contractors' payroll and 2 per cent payments for labour-only subcontractors, became entirely employer-determined and severed from collective bargaining processes (Clarke 1992a). This compares with the 2 per cent FRG training levy, determined through collective agreement and administered by a common social fund (Streeck and Hilbert 1991).

In Britain rather than upskilling the workforce, efforts continued to be concentrated on certifying existing and traditional skills without significant investment in further training. Initial VET remained largely trade-based, narrow and relatively untheoretical, with trainees entering a particular trade, not the industry as a whole. The training system, too, remained dependent on the individual employer to take on trainees, based on the assumption of a stable, directly-employed workforce able to guide, supervise and monitor trainees' progress, and – increasingly – on the 'learning by example' principle and, by the 1990s, a return to day rather than block release. As the number of placements with employers declined, the Further Education (FE) sector stepped in to become the main route to obtaining training in construction, though at a lower level than is produced through the German system (Steedman 1998).

Differences from Germany

The most notable characteristic of labour in the British construction industry is its continued strong trade basis, with large areas unrecognised as 'skilled', including groundwork, concreting, paving, and machine operation and cladding. These are all 'skilled' areas in Germany but to all intents and purposes regarded and paid as labourer's work in Britain, despite the introduction of a General Construction Operative National Vocational Qualification (NVQ). Indeed in Britain, unlike most European countries, qualifications are not linked directly either to the wage structure or to VET. Despite changes to the collective agreement to introduce a new "skilled operative" status on sites, divided into four grades, this has continued to be distinguished from the "craft operative" and the "general operative" (CIJC

2003). In Germany, in contrast, few activities remain outside the realm of the skilled worker, so that the labourer is fast becoming marginal on building sites. The key means for skills to be socially recognised is through the collectively agreed wage, with eight grades each linked to a particular skill and training level; if new skills with appropriate training are recognised, this is reflected in the wage. The social indirect wage forms a critical part of the wage structure and payment is based on hours worked and qualifications rather than output.

This lack of recognition in Britain of a large area of building activity contributed to a low proportion of qualified construction workers by the 1990s, with only 36 per cent achieving NVQ Level 3 compared with 83 per cent in the German workforce (Richter 1998). While a multitude of NVQs had been developed for construction (the result of lobbying by numerous trade and employer associations), by the mid-1990s about 86 per cent of trainees were still to be found in traditional trades, for which FE Colleges largely catered – bricklaying, carpentry and joinery, painting and decorating, plastering, plumbing and electrical work – though only 59 per cent of the workforce were employed in these (i.e. the new NVQs were not taken up, or only to a very minor extent).

In Germany, by contrast, all construction activity from the 1980s onwards came to be covered by 15 occupations, each encompassing a wide range of skills and grouped into three areas for training purposes: building, civil engineering and finishing (Clarke and Wall 1996). In Britain, the trades do not exhibit the same universality and remain synonymous with contractual divisions, especially those defined through labour-only subcontracting.

By the turn of the millennium, the proportion of carpentry and bricklaying trainees to those employed in these trades was 4 per cent in Britain, compared with 22 per cent in Germany (IMI 2002). In the other countries, with systematic training and upskilling, the construction workforce became almost entirely skilled. In Britain, in contrast deskilling occurred as the number of first-year entrants to construction training fell by 18 per cent during the 1990s (CITB 2002); over the same period in Germany trainee numbers rose by 65 per cent, even though the overall number of skilled workers fell. In terms of the depth and extent of training, Britain also presents a contrast with the majority of those training for a trade undertaking less than two years training to NVQ level 2, whilst in Germany training is three years. At the same time in Britain areas of abstract knowledge, such as mathematics, were significantly reduced in the curriculum (Steedman 1992). The system came to be marked by a separation of theoretical and practical knowledge and a general lack of underpinning skills essential to transferability, whether to higher progression to technical levels or for easy adaptation to new methods. In Germany, in contrast, the focus has been more on education and on imparting transferable skills, so the VET system is broader, integrating theoretical and practical work-based elements, and consisting of more general educational elements, such as languages – geared essentially to developing potential or in Biernacki's (1995) terms, 'power' of labour.

The tripartite industry-wide organisation of VET in Germany has also made for less dependence on the single employer. The trainee is a trainee to the industry and VET divided into three locations, each with a distinct role, the college, the workshop and the site (Clarke and Wall 1996). The training workshops provide a broad introduction to practical activities and are industry-run by the social partners and funded through levy contributions, and in Germany state-supported to deliver training in the most advanced techniques. The role of site experience is seen as training 'for the market'. Indeed, the long-term development of the VET system has been a gradual move away from 'learning on the job', given the increasing difficulties of this, especially when components and machinery used are more valuable, firms may be very specialist and the skills required more abstract (associated with planning, programming, calculating, setting out and measuring) – all best taught in the classroom or workshop.

The continued craft nature of the division of labour in Britain is manifest also on site. Interfaces, for instance between the door and the wall, tend to be defined by trade boundaries, often employed through labour-only subcontractors, whether for brickwork, carpentry or steel fixing. This makes for significant problems of coordination and high levels of supervision to ensure quality of output. In contrast, in Germany, though there are many specialist subcontractors employed especially on the finishing work, labour-only subcontracting remains illegal and interfaces are not defined by trade. Skilled operatives are trained in advanced techniques and to a definite standard, and are expected to be responsible for planning, carrying out and controlling the work they undertake, so require little supervision. The degree of prefabrication and levels of mechanisation are much higher than in the British case, whilst the level of subcontracting is lower (Clarke and Herrmann 2004). And the construction process in Germany has also become less and less manual, with the proportion of non-manual workers in the workforce more than doubling in the last third of the twentieth century, whereas in Britain it has remained the same (DTI 2004; *Die Deutsche Bauindustrie* 2004).

Conclusions

The decline and erosion of industrial democracy in Britain provide one explanation for the increased disparity with continental construction labour processes since the 1960s and 1970s. In Germany the system of social partnership, and with it the role of the unions, has developed significantly with the extension of the social wage and training provision and the different industry funds under joint committees of unions and employers operating at local, regional and national levels. The levy-based training organisations are tripartite with training committees at regional and local levels composed of educationalists, employer and employee representatives. Most important of all, collective agreements made at national level are statutory, so apply to the majority of the workforce. At firm level, too, their implementation is overseen by workers' councils.

We could continue to search for explanations of differences within a comparative framework *ad infinitum*. Perhaps a more pertinent question is whether labour processes such as those in the construction sector are specific to particular socio-legal frameworks or do they, in the course of their development, transform, transcend and even dissolve these? And, if they are assumed to be specific, surely they – and therefore also the social-legal framework – cannot be regarded as homogeneous, without contradictions and disparities? Even in Britain, sharp disparities exist between employment and training conditions in the mechanical and electrical, civil engineering and building sectors, each with rather separate collective agreements, VET traditions and occupational profiles.

All the indications are, as illustrated in **Table 3.1**, that we are now in a new stage of development, marked by the growing importance of qualifications for entry into the labour market and of full-time college-based VET schemes, by a more individualised employment relation rather than attachment to a single employer, graded skill levels, and a change in social partnership, as unions and employers associations assume a different role and structure. These tendencies are apparent, to a greater or lesser extent, in all the leading European countries considered here.

If we consider developments in these terms rather than on a national basis the evident disparities are rather between:

- a socially regulated wage versus an unregulated individual one;
- collective industrial relations versus individualised employment relations;
- comprehensive, industry-wide VET programmes versus trade-based training;
- and qualification-based versus performance/output-based labour markets.

Such disparities exist to a greater or lesser extent in all countries and overall echo Biernacki's (1995) distinction between labour power and 'embodied' labour as well as Marsden's (1999) 'training' and 'production' approaches. For Marsden the 'training approach' is one where investment in training is provided by collective industry-related associations of employers and employees together with the state and which exhibits a more regulated approach to the labour market. This approach underpins 'occupational labour markets', which might be described as 'qualification-based' in the sense that entry is dependent to a large extent on VET and qualifications. It contrasts with a 'production approach', associated with the trade or craft labour markets and with the traditional apprenticeship, bound to a single employer so that skills tended to be firm-specific and training largely dependent on the individual employer and on on-the-job learning.

Surely, for the construction industry in Britain, as in the rest of Europe, the historical tendency is to move from a trade-based to an occupational labour market, as evident in the increasing reliance on college-based VET systems, on qualifications linked to wage grades, and on broader occupational profiles. If this is the case, and we are entering a stage of development where qualifications rather than 'who you know' are critical for entry, this has far-reaching implications for the industry in Britain, bringing it closer to other systems elsewhere. The construction labour markets of Germany and the Netherlands already resemble occupational labour markets, defined by Marsden (1999, 2007) as institutionally regulated, related to a person's skill and certified qualifications rather than just to performance in the workplace, and usually collectively and industrially organised. For this to be the case in Britain, a more comprehensive VET system and a more regulated labour market are required, involving clear negotiation and recognition of a restricted range of occupations covering all construction activities and contributing to a clearly defined division of labour.

Notes

1   The organisations and acronyms are defined in the following discussion
2   National Archives Lab 8/518 The Education Committee consisted of: E. D. Simon (chairman), G. Burt, R. Coppock, L. Fawcett, Hindley, Mr Laing, Loudon, McTaggart, Thomas, Pitt, G. D. H. Cole and representatives of various government departments.
3   National Archives Lab 8/518. Advance Copy of Part III of G. D. H. Cole's Report, p. 4.
4   The regional variations were considerable. In Scotland, almost 50% of local authority houses built between 1945-55 were constructed using non-traditional methods. In London the figure was just 8%.
5   i.e. labourers receive 80% of the rate of craftsmen.
6   With contributions every week.
7   Workers using particular types of excavators could earn 2 shillings 3d per hour plus rate, a figure that put them well ahead of craft workers (Hilton 1968:145).
8   AUBTW and UCATT activist, Bert Lowe's (1996) memoir contains several mentions of successful struggles against lump labour: pp64, 86-88.
9   Craft basic pay rose from £20 per week to £26 per week, labourers basic pay rose from £17 to £22.20.
10 Preface to BATC Final Report (1956) by the chairman F.W. Leggett.
11 This was the pilot version of the Standard Scheme of Training, which was subsumed into the Youth Training Scheme (YTS) in 1982.
12 i.e. Being paid an extra month's salary in December to help over the Christmas period.

References

Allen, V. L. (1952) Incentives in the building industry, *Economic Journal*, Vol. 62, No. 247, September, 595–608.

Austrin, T. (1978) *Industrial Relations in the Construction Industry: Some sociological considerations on wage contracts and trade unionism (1919–1973),* University of Bristol, PhD thesis.

Austrin, T. (1980) The 'lump' in the UK construction industry, Nichols T. (ed) *Capital and Labour,* London.

BATC (Building Apprenticeship and Training Council) (1944) *The Second Report of the BATC,* London: HMSO.

Bercusson, B. (1978) *Fair Wages Resolutions: studies in labour and social law,* London: Mansell.

Biernacki, R. (1995) *The Fabrication of Labour: Germany and Britain 1640–1914,* Berkeley: University of California Press.

Bowley, M. (1966) *The British Building Industry,* Cambridge: Cambridge University Press.

Brockmann, M., Clarke, L., Hanf, G., Méhaut, P., Westerhuis, A., Winch, C. (2011) *Knowledge, Skills, Competence in the European Labour Market: What's in a qualfication?* Oxford: Routledge.

Building Workers Charter (1972) Vol. 2, No. 3.

Burchell S. (1980) Training: an analysis of the crisis, *Proceedings of the Bartlett International Summer School 1979,* University College London.

Campinos-Dubernet, M. and Grando, J-M. (1991) *L'analyse sectorielle comparative: questions, méthodes, résultats* in CEREQ *Europe et chantiers,* Paris: Plan Construction et Architecture.

Campinos-Dubernet, M. and Grando, J-M. (1992) Construction, Constructions: a cross-national comparison in *The Production of the Built Environment, Proceedings of the 11 Bartlett International Summer School,* Paris 1991, London: University College London.

CEREQ (*Centre d'études et des recherches sure les qualifications*) (1991) *Europe et chantiers: le BTP en Europe: structure industrielles et marché du travail,* Actes du collogue, 28–29 September 1988, Paris: Plan Construction et Architecture.

Clarke, L. (1992a) *The Building Labour Process: problems of skills, training and employment in the British construction industry in the 1980s,* Chartered Institute of Building, Occasional Paper No. 50.

Clarke, L. (1992b) Training Provision and Wage Form: the example of the construction labour process in Europe, paper to the 14th International Working Party on Labour Market Segmentation, Cambridge.

Clarke, L. (1999) The changing structure and significance of apprenticeship with special reference to construction in Ainly, D. and Rainbird, H. (eds) *Apprenticeship: towards a new paradigm of learning,* Kogan Page, 25–40.

Clarke, L. and Herrmann, G. (2004) Cost versus production: disparities in social housing construction in Britain and Germany, *Construction Management and Economics,* June, 22, 521–32.

Clarke, L. and Wall, C. (1996) *Skills and the Construction Process: a comparative study of vocational training and quality in social housebuilding,* Bristol: Policy Press.

Clarke, L. and Wall, C. (2011) Skilled versus qualified labour: the exclusion of women from the construction industry, Davis, M. (ed), *Class and Gender in British Labour History,* Merlin Press.

Cole, G. D. H. (1918) *The Payment of Wages: a study in payment by results under the wage-system,* London: Allen & Unwin.

Cole, G. D. H. (1945) *Building and Planning*, London: Cassell.

Collingwood, R. G. (1961) *The Idea of History*, Oxford: Oxford University Press.

Connelly, T. J. (1960) *The Woodworkers 1860–1960*, London: ASW.

Construction Industry Joint Council (CIJC; 2003) *Working Rule Agreement for the Construction Industry*, London.

Construction Industry Training Board (CITB) (1966) *Report and Statement of Accounts*, London: HMSO.

Construction Industry Training Board (CITB) (1967) *Report and Statement of Accounts*, London: HMSO.

Construction Industry Training Board (CITB) (1969) *A Plan of Training for Operative Skills in the Construction Industry*, London: HMSO.

Construction Industry Training Board (CITB) (2002) *Trainee Numbers Survey*, Bircham Newton.

Cotgrove, S. F. (1958) *Technical Education and Social Change*, London: George Allen and Unwin.

Deißinger, T. (1998) *Beruflichkeit als 'organisierendes Prinzip' der deutschen Berufsaussbildung*, Markt Schwaben: Eusl.

Department of Trade and Industry (DTI) (2004) *Construction Statistics Annual 2003*, London.

Die deutsche Bauindustrie (2004) *Baustatistisches Jahrbuch 2003*, Frankfurt am Main.

Druker, J. (1991) Wage forms in construction: regional differences in the development of wage forms, Clarke L. (ed) *The Development of Wage Forms in the European Construction Industry*, Proceedings of a Conference 4–6 October, Dortmund: Fachhochschule Dortmund.

Druker, J. (1992) *National Bargaining in the Building Industry: how and why does it survive?* Unpublished paper.

EFBWW (European Federation of Building and Woodworkers) (1991) *Conditions of Employment in the European Construction Industry*, Brussels: EFBWW

Girdwood Report (1948) *The Cost of Housebuilding: First Report of the Committee of Engineering appointed by the Minister of Health*, London: HMSO.

Greinert, W-D. (2007) The German philosophy of vocational education, Clarke, L. and Winch, C. (2007) *Vocational Education: international approaches, developments and systems*, London: Routledge.

Handy, L. J. (1971) *Wage determination in the Construction Industry: an analysis of post-war trends*, Department of Applied Economics, University of Cambridge.

Hanf, G. (2007) Under American Influence? The making of modern German training in large Berlin enterprises at the beginning of the twentieth century, Clarke, L. and Winch, C. (2007) *Vocational Education: International approaches, developments and systems*, London: Routledge

Hilton, W. S. (1963) *Foes to Tyranny: A History of the AUBTW*, London: AUBTW.

Hilton, W. S. (1968) *Industrial Relations in Construction*, London: Pergamon.

Howell, G. (1877) Trade unions, apprentices and technical education, *Contemporary Review*, 30, 833–57.

IMI (Innovative Manufacturing Initiative) (2002) *Standardisation and Skills, final report*, University of Westminster.

Janssen, J. (1992) The Building Labour Process in FRG: a historic definition, *The Production of the Built Environment: Proceedings of the Bartlett International Summer School 11*, London: University College London.

Jeanes, R. E. (1966) *Building Operatives Work Vols. 1 and 2*, Ministry of Technology, Building Research Station, London: HMSO .

Lanove, D. (1990) *The Wage Cost in the European Construction Industry*, Brussels: Fédération Nationale de la Construction de Belgique.

Lee, D. (1979) Craft union and the force of tradition, *British Journal of Industrial Relations*, 17(1) 34–49.

Lowe, B. (1996) *Anchorman*, Stevenage: Bert Lowe.

Margirier, G. (1991) Le secteur du bâtiment et des travaux publics dans la crise: comparaison France, RFA, Italie, Royaume-Uni, CEREQ *Europe et chantiers*, Paris: *Plan Construction et Architecture.*

Marsden, D. (1999) *Theory of Employment Systems: microfoundations of societal diversity*, Oxford: Oxford University Press.

Marsden, D. (2007) Labour market segmentation in Britain: the decline of occupational labour markets and the spread of 'entry tournaments', *Économies et Sociétés*, 28/6, 965–98.

Ministry of Works (1950) *Wages, Earnings and Negotiating Machinery in the Building Industry since 1886*, Economic Research Section.

National Board for Prices and Incomes (NBPI) (1968) *Pay and Conditions in the Building Industry*, London: HMSO.

National Economic Development Office (NEDO) (1991) *Restructure to Win*, Construction Industry Sector Group, January.

New Builders Leader (1941) *Problems of the Building Industry*, with Marx House, London: Lawrence and Wishart.

OECD (1964) *Vocational Training in the Enterprise in the Context of Industrial Change*, report of visit by German joint team, 2–7 November.

Pellegrini, C. (1990) *Collective Bargaining in the Construction Industry: wages, hours and vocational training in Belgium, the Federal Republic of Germany, France, Italy, Spain and the United Kingdom*, Luxembourg: Office for the Official Publications of the European Communities.

Perry, P. J. C. (1976) *The Evolution of British Manpower Policy: From the Statute of Artificers to the Industrial Training Act 1964*, Portsmouth: Grosvenor Press.

Phelps Brown (1968) *Report of the Committee of Inquiry under Professor E. H. Phelps Brown into Certain Matters concerning Labour in Building and Civil Engineering*, London: HMSO.

Postgate, R.W. (1923) *The Builders' History*, London: National Federation of Building Trade Operatives.

Price, R. (1980) *Masters, Unions and Men: work control in building and the rise of labour 1830–1914*, Cambridge: Cambridge University Press.

Ribeill, G. (1991) Une forme archaique d'organisaiton sociale du travail à l'épreuve de l conjuncture économique: le tâcheronat, Clarke L., *The Development of Wage Forms.*

Richter, A. (1998) Qualifications in the German construction industry: stocks, flows and comparisons with the British construction sector, *Construction Management and Economics* 16(5) 581–92.

Schudlich, E. (1991) Ideas regarding the historical relation between wage payment systems and work organisation, in Clarke, L. (ed) *The Development of Wage Forms in the European Construction Industry*, Proceedings of International Conference, 4–6 October 1990, Fachhochschule Dortmund.

Simon Committee (1943) *Report on Training for the Building Industry*, London: HMSO.

Steedman, H. C. (1992) Mathematics in vocational youth training for the building trades in Britain, France and Germany, *NIESR Discussion Paper 9*, London.

Steedman, H. (1998) A decade of skill formation in Britain and Germany, *Journal of Education and Work*, 11(1) 77–94.

Streeck, W. and Hilbert, J. (1991) Organised interests and vocational training in the West German construction industry, Rainbird and Syben (eds) *Restructuring a Traditional Industry*, Oxford: Berg.

Tallard, M. (1991a) Métier et branche, elements indissociables de la régulation dans le BTP? Une comparaison France/Grande-Bretagne, CEREQ *Europe et chantiers*, Paris: Plan Construction et Architecture.

Tallard, M. (1991b) Context and limits of policies of social innovation in small and medium-sized construction firms in France, Rainbird and Syben (eds) *Restructuring a Traditional Industry*, Oxford: Berg.

University of Manchester Institute of Science and Technology (UMIST) *Building Industry Wage Structure*, IPC, Building and Contract Journals Ltd.

Weis, U. (1991) Processes of change in historical perspective: technology and technique in the construction process 1919–1933 in Germany, Rainbird and Syben (eds) *Restructuring a Traditional Industry,* Oxford: Berg.

White Paper (1943) *Training for the Building Industry,* Cmd 6428, London: HMSO

White Paper (1956) *Technical Education,* Cmd 0903, London: HMSO.

Winch, G. (1992) *Self-employment and the Labour Process: the Example of Construction,* paper presented at 10th International Labour Process Conference, Aston.

Wood, L. (1979) *UCATT: A Union to Build*, London: Lawrence and Wishart.

4 Human resource development in construction: moving beyond alignment with organisational strategy

Paul Chan and Mick Marchington

Introduction

Employers often claim that people are their greatest asset. It might be expected therefore that organisations would take measures to nurture their human resources in order to survive in a competitive marketplace (Pfeffer, 1998). Indeed, there is growing recognition that the organisation and enhancement of human capability, or human resource development (HRD), play an integral part in the survival and sustainability of organisations (see Swart *et al.*, 2005; Barney, 2007; Wright *et al.*, 2007). Writings about HRD have increasingly emphasised the need for HRD practices (e.g. training and development) to align with organisational strategy. So, HRD choices are made to ensure strategic fit, both in terms of internal operations and the external environment, to secure organisational success (see Huselid, 1995; Wood, 1999; Paauwe, 2004) and sustain competitive advantage (Barney, 2007; Wright *et al.*, 2007).

Despite rhetorical claims about the benefits of HRD, the reality of HRD is far from straightforward. Debates have developed over the disconnections that exist between the rhetoric and reality of HRD (see e.g. Keenoy, 1999; Garavan *et al.*, 1999; McCracken and Wallace, 2000; Sisson and Storey, 2000; Legge, 2005; Sambrook, 2009). Grugulis *et al.* (2004) suggested that contemporary claims of investing in the workforce through upskilling initiatives do not always translate to real benefits of wage premiums to employees (see also Nyhan *et al.*, 2004), as employers continue to abdicate from their responsibility to train and develop workers. Modern work organisation remains doggedly Taylorist and many employers often postpone or abandon training programmes in order to meet production schedules (see e.g. Thursfield, 2001; Grugulis, 2007). Scholars have also questioned whether HRD initiatives such as 'Investors in People' actually deliver any value to organisations (see Bell *et al.*, 2002; Grugulis and Bevitt, 2002, and Smith, 2009).

The divergence between theoretical aspirations and practised reality of HRD has been well-rehearsed in the literature. Distinctions have also been made between 'hard' and 'soft' HRM (see e.g. Druker and White, 1996; Druker *et al.*, 1996; Truss *et al.*, 1997), where the former treats workers simply as commoditised resources to meet the performative needs of the organisation, and the latter emphasises the enabling of human capabilities to flourish in organisational life. The reality of HRD also mirrors more of the former than the latter. Many employers tend to adopt the 'low-road' route to employee development typified by a low-skill–low-wage strategy aimed at controlling the cost of labour, as opposed to the 'high-road' route of investing in training and development to secure strong employee commitment and competitive advantage on the basis of quality (see e.g. Youndt *et al.*, 1996; Wood and DeMenezes, 1998; Blyton and Turnbull, 2004; Orlitzky and Frenkel, 2005). Certainly in the context of construction companies, scholars have suggested that HRD is rarely seen as a core business activity (see Beckingsdale and Dulaimi, 1997; Forde and MacKenzie, 2007; Chan and Dainty, 2007).

In this chapter, we seek to build on the knowledge of disparities between the rhetoric and reality of HRD in construction by exploring the reasons for this gap. Our central argument is that aligning HRD practices with organisational strategy is insufficient in itself to ensure that benefits of HRD are realised. To fulfil the aspirations of HRD, companies in the construction industry would need to consider three fundamental challenges. Firstly, companies must recognise the tensions that exist between managers and workers when constructing the HRD agenda in organisations. Such asymmetrical power relationship between managers and workers can limit the possibilities of enacting successful HRD practices. Thus, there is a need to consider the interpersonal dynamics between managers and workers in organisations. Secondly, construction projects are seldom delivered by a single organisation. Rather, the project-based nature of the construction industry is typified by temporary coalitions of organisational partners (Groák, 1994). This implies that a straitjacket approach to linking HRD with strategy *within* an organisation is inadequate because this ignores the inter-organisational relations that matter in construction. Yet, there is scant attention given to how HRD is considered *across* organisations in project-based construction (Dainty and Chan, 2011). Thirdly, effective HRD practices demand the need for organisations to look outside their boundaries to consider institutionalised contexts (see e.g. DiMaggio and Powell, 1983) within which organisations (often unstable, incoherent and contradictory) operate. For instance, when engaging with the vocational education and training (VET) system for the development of human resources, there is a need for employers to consider the wider political economy of skills formation (see e.g. Chan and Moehler, 2008). Thus, HRD practices are influenced by interpersonal, inter-organisational and institutional dynamics that transcend the simplified explanation of strategic alignment that predominates the literature. The intention of this chapter is therefore to clarify levels of organisational analysis. Following Marchington and Vincent (2004), the proposition is that interpersonal, inter-organisational and institutional layers of HRD will need to be scrutinised more carefully to understand how the aspirations of HRD can be realised in construction.

Our arguments are organised in three main sections. Firstly, ideas about HRD are reviewed. This traces the historical development of HRD from the chaotic and informal approaches to training associated with the tools of the trade in feudal times, to more formal approaches of VET in the present. The review also highlights contemporary interest in knowledge work, and how this has renewed the belief that people are critical to securing competitive advantage for organisations (Barney, 1991; Grant, 1996). In the second section, this notion of strategic alignment between HRD and organisational performance and competitiveness is problematised in the context of interpersonal, inter-organisational, and institutional relations in the construction industry. Here, the argument is made for policy, research and practice endeavours to consider the power struggles that are inevitable when constructing HRD approaches across a plural landscape of stakeholders (including *inter alia* workers, managers, employers, supply chain organisations, institutional actors etc.). Finally, a discussion on the potential implications for policy, research and practice is presented in the third section. The chapter is punctuated with case studies from the UK construction industry to illustrate the challenges of aligning HRD across interpersonal, inter-organisational and institutional dimensions, which in turn questions current HRD practices in construction.

Ideas about HRD

Moving from functional instrumentalism to strategic alignment

Defining the boundaries of what constitutes HRD is a process fraught with frustration, vagaries and confusion (McGoldrick *et al.*, 2001). Numerous authors have noted that HRD is a socially-constructed, discursive concept (e.g. Garavan *et al.*, 1999; Vince, 2005; Sambrook, 2007; 2009); and so, as with most social constructs, one can examine HRD with multiple theoretical lenses. Some scholars have gone even further to suggest that the quest to define HRD in a comprehensive way remains futile since the field is an emergent and evolving one (Lee, 2001; Swanson, 2007). Nonetheless, as **Table 4.1** shows, there are common features that have been identified by a range of authors in the field. For instance, Harrison (1997; cf. Garavan *et al.*, 1999: 171) considered HRD to include '[...] the provision of training, development and education activities designed to enhance the utilisation of human resources within the organisation and contribute to the achievement of explicit corporate and business strategies.' Swanson and Holton (2009) also surmised that HRD is '[...] a process of developing and unleashing expertise for the purpose of improving individual, team, work process, and organizational system performance (p. 4).' Therefore, critical features include training and development activities alongside issues concerning wider organisational learning and development activities that cut across all levels of work organisation to meet strategic objectives (see Armstrong, 2003).

Table 4.1 Critical features of HRD espoused in recent literature

Author	Critical features of HRD espoused in the literature
Garavan et al. (1999: 169)	• HRD is intrinsically aligned to the identification of core competencies across every organisational level so as to correspond with organisational strategy for sustaining competitive advantage now and for the future. • HRD is an investment in human resources to accrue benefits associated with organisational and individual learning. • HRD is a social and discursive construct that views the employee in a holistic sense. • HRD is connected to, and mutually reinforces, other HR strategies.
Lee (2003: 21)	• HRD is connected with the study of the system and relationships within an organisation; its understanding is facilitated by a multiplicity of representations. • HRD is about agency in a pluralistic context, aimed at searching for dynamic structures and understanding possibilities. • HRD is an emergent voyage of 'doing' and 'becoming'.
Swart *et al.* (2005)	• HRD is part of an organisational core competence, intrinsically linked to the high performance agenda. • HRD involves enabling organisational structures to instil a shared purpose, learning and development and employee participation and involvement. • HRD is connected with changing the way an organisation behaves and the establishment of organisational values over time.
Russ (2005: 26–27)	• HRD requires the negotiation of what learning and change mean and involve within organisations. • HRD occurs within a pluralistic context where different power interests exist and compete. • HRD requires practitioners to make sense of, and transform the various practices that emerge within specific situations of learning and development.
O'Donnell *et al.* (2006)	• HRD is instrumental in encouraging employees to drive organisational performance. • HRD benefits are contributed by, and accrued to employees. • HRD is distinct from, but pragmatically connected to, HRM. • HRD is framed in a pluralistic context, aimed at creating positive organisational and learning cultures to instil norms and values.

Author	Critical features of HRD espoused in the literature
Wright *et al.* (2007: 80)	• HRD focuses on developing higher levels of skills or seeking better alignment between the skills represented in the firm (human capital pool) and those required by its strategic intent. • HRD is also about aligning individual behaviour to achieve the goals of the firm. • HRD is a dynamic process that requires the understanding of multiple practices within the repertoire of people management systems that impact on employees.
Sambrook (2009: 63)	• HRD is related to training and development, employee development, organisational learning and critical management studies. • HRD is framed to differentiate between, yet combine, strategic and business-oriented learning and development activities from conventional training and development. HRD is a social and discursive construct to connect ways of thinking, talking about and practising HRD. • HRD helps individuals achieve their own aspirations whilst transforming the socio-political structures in which they exist.
Swanson and Holton (2009: 10)	• HRD is connected to the fate of any organisation. Organisations rely on human expertise to succeed in a changeable and turbulent environment. • HRD tools and processes are deployed to seek positive and fair outcomes, whilst ensuring short-term and long-term gains for individuals and the organisation. • HRD practitioners work across individual, group, work process and organisational boundaries.

In this section we present a brief historical overview of HRD to show how the framing of HRD–at least in theory–has shifted away from direct control of the attributes (e.g. skills, behaviours) of the workforce, to one that seeks alignment with organisational strategy in a modern economy based around knowledge work through more indirect and subtle means. Two main issues emerged in this transformation. First, as HRD develops to take on a strategic turn, its activities become more formalised and organised. Second, the focus shifts towards aligning human behaviour and expertise to organisational systems and structures. In seeking closer integration between employment structure and agency, greater emphasis is placed on the performance agenda. This link between HRD and organisational performance forms the crux of much writing in this area.

Early origins of HRD

Clarke (1999) traced the early origins of HRD in construction by explaining the changing structure and significance of the apprenticeship system in the UK construction industry. So, indentured apprenticeships (i.e. where apprentices are bound to an individual master) formed the basis of the guild apprenticeship system up until much of the sixteenth century. The guild apprenticeship system created a legal and paternal bonding between master and apprentice, characterised by a master–servant relationship where the master provided the apprentice with the daily essentials of food and lodging, and the apprentice would provide the master with his labour in terms of cleaning, fetching and carrying the tools. Learning took place over several years, and was undertaken mainly through experiential modes of learning-by-watching and learning-by-doing. This progressed on to a more regulated, statutory apprenticeship system in the period from the middle of the sixteenth century to early nineteenth century, where the industry saw the development of wage labour and a formalisation of a range of trades and occupations. The ascendency of the trade union movement saw the development of the collectively-bargained apprenticeship, which formalised the organisation of labour, including such matters as training and development.

Rise of the human capital perspective

Over time, the development of human ability has centred on three critical dimensions, including education to enhance the intellectual capability of the individual, training to improve the physical dexterity and handling of the tools and materials of the trade, and experiential modes of learning-by-watching and learning-by-doing. These have taken on more formal methods as time has progressed, and form the basis of functional approaches to learning and development (see Buckley and Caple, 2004). Clarke and Winch (2004), in arguing the case for a comprehensive system of VET for the construction industry, stress that these three dimensions are essential for the development of sound vocational practice. They suggest the development of human ability and knowledge in an applied context such as construction necessitates a combination of theoretical knowledge transfer, work-based training opportunities, and experiential learning. However, developments after the Second World War posed some challenges to this fundamental model of HRD.

In 1964, Becker published his seminal work on *Human Capital*, which laid out the arguments for investing in HRD so as to reap rational economic benefits for individuals. Policymakers found the concept persuasive when framing policies surrounding a range of issues, from education to economic development to immigration. But the trouble with a rational perspective such as that presented by Becker lies in defining what a skill actually is. As Clarke (1992) wrote about the construction labour process:

[...] whilst training creates skills, these skills have different values for the worker who owns, sells, employs and attempts to conserve them than for the builder (employer) who buys and consumes them [...] Under a capitalist mode of production [...] the determination of training provision is only possible through an analysis of changes in production and in the social relations regulating the labour process (p. 6).

For Clarke, skill is a socially constructed concept, deeply embedded within the socio-political context in which it is afforded. In order for Becker's (1964) rational economic model to work, there needs to be a quantifiable means of measuring skills. Of course, the best proxy lies in qualification levels (see Stasz, 2001; Grugulis *et al.*, 2004), although this is not an uncontested area of educational policymaking (see Clarke and Winch, 2006; Brockmann *et al.*, 2009).

Admittedly, two consequences of Becker's (1964) human capital perspective can be observed. Firstly, by constructing human capital (skills) in terms of qualification levels, there is a convincing case for individuals to pursue higher, tertiary levels of education. One phenomenon that corresponds with this trend is the increasing professionalisation seen since the post-War era, which gave rise to the professional manager in the construction industry, increasingly educated through University degree programmes. This replaced the time-served tradesperson who used to advance through the ranks conventionally by amassing technical experience and knowledge through crafts training and the delivery of building work (see e.g. Langford and Hughes, 2009). On the one hand, the scope of construction has expanded beyond traditional trade boundaries to include a sizeable proportion of professional services such as design work and professional consultancy (Pearce, 2003). At the same time, there is also a widening gap between the status of professional workers and craft labour (Clarke and Herrmann, 2007), and a shifting emphasis that places more credence on quantitative measurement rather than quality of skills, which in turn result in greater concentration on theoretical knowledge as opposed to vocational education and training (Clarke, 2006).

Secondly, the rational economic approach of the human capital perspective probably explains the positivistic paradigm within which HRD studies had been framed (Valentin, 2006). HRD models (see e.g. Holton, 1996, and; Holton *et al.*, 2000) provide a mechanistic approach to capture the outcomes of learning accrued from a combination of training, experience and personal motivation to learn in terms of effects on individual performance, which is then aggregated to organisational performance. The learning model developed by Bee and Bee (2003) portrays a cycle of identifying, specifying, translating, planning and evaluating learning needs of individual workers vis-à-vis the contribution to business needs. Winterton (2007) also asserted:

According to the conventional wisdom of 'nuts and bolts' personnel management, having established personnel requirements (taking into account labor turnover, retirements, sales forecasts, and the impact of technological changes on productivity), recruitment, selection, and training follow as a linear trilogy. A workforce with the requisite skills is the logical end result, enabling the personnel team to focus on appraisal, remuneration, and motivation until the next round of 'manpower planning' [...] modern HRM might emphasize the need for continuous training, and development to maintain the dynamic capabilities supporting organizational strategy and make endless caveats about choices to be made between recruitment, training and outsourcing (p. 324).

It seems, therefore, that the sole object of HRD has been driven by the agenda of organisational performance and that the individual worker, and the benefits associated with individual learning remain secondary in the process (see e.g. Sisson and Storey, 2000). As Loosemore *et al.* (2003) pointed out:

The rationale behind investing in HRD is that investing in people in the right ways will ensure that they continue to contribute to the direction in which the business wants to go. Failing to address HRD needs inevitably leads to the reopening of skills gaps (p. 94).

Alignment with organisational strategy and the agenda of the knowledge economy

Boxall *et al.* (2007), however, considered HRM to be instrumental in renewing top management's interest in bringing 'people issues' to the fore. By framing competitive strategies in terms of the resource-based view of the firm, the attention of senior management can now be re-directed to reflecting upon the core competencies (Hamel and Prahalad, 1994) and dynamic capabilities (Teece *et al.*, 1997) necessary to maintain organisational agility. In turn, competitive advantage is yielded by developing rare, valuable, inimitable and non-substitutable human resources (Barney, 1991; 2007). As Boxall *et al.* (2007) observed:

[...] writers in general and strategic management continue to downplay the importance of work organization and people management. To be sure, resource-based theory has reawakened the human side of strategy [...] HRM is central to developing the skills and attitudes which drive good execution. This in itself is enormously important but, more than this, the contribution of HRM is dynamic: it either helps to foster the kind of culture in which clever strategies are conceived and reworked over time or, if handled badly, it hinders the dynamic capability of the firm (p. 8).

Thus, within the strategic HRM context, there is a strong sense that the development of human resources should constantly align with organisational strategy (see also the critical features of HRD depicted in Table 4.1). Contemporarily, efforts to align HRD practices are couched within the rhetoric developed around the notion of the knowledge economy, where a firm's competitiveness is sustained by harnessing its knowledge assets (Grant, 1996) to move up the value chain (Porter and Ketels, 2003). As Littler and Innes (2003) noted, 'Underlying the knowledge capitalism paradigm was the argument that "skill" had become decentred in the production process (p. 75).' They cited Aronowitz and DiFazio's (1994) assertion that:

> The twentieth century [...] is marked by the displacement of skill by knowledge. It is the knowledge component–the conceptual, the theoretical–that is now the basis for the scientific, technological and social relations of production (p. 95; cf. Litter and Innes, ibid.).

This reinforces Clarke's (2006) reflection on the shifting focus towards theoretical knowledge endorsed by proponents of the human capital perspective. Consequently, greater emphasis has been placed on the essence of learning (Stata, 1989), where the aspirations of developing learning organisations (Senge, 1990) coincide with policy attempts to encourage lifelong learning approaches in the individual (see e.g. Commission of the European Communities, 2000; 2001). Commentators have also noted the optimism of such knowledge-based approach in the study of construction organisations (see Kululanga *et al.*, 2001, and; Chan *et al.*, 2005).

In summary, the acquisition and formation of skills and knowledge have thus broadened from the individual perspectives of learning-by-doing and learning-by-watching to include greater participation by state and corporate actors to systematise HRD through more institutionalised forms of education and training. The dominant view of modern HRD is typically mechanistic and utilitarian. The emphasis has been placed on linear approaches that habitually consider the identification of skills and knowledge gaps expressed in quantifiable and commoditised ways, the implementation of learning, education and training plans, and the evaluation of HRD activities. The mission of HRD appears to be preoccupied with the agenda of alignment, where the pursuit of individual learning is intimately connected with the fulfilment of the strategic objectives defined by their organisational masters, and governmental policy aims of encouraging a society that partakes in lifelong learning. The focus is narrowly based on performance, in which individual learning should contribute to enhancing the bottom line for business and the wider economy. Such emphasis on aligning HRD and the strategic aims of the organisation is illustrated in the two case studies below.

Case Study 4.1

Family matters in the John Doyle Group

This case study has been adapted from Cooper and Chan (2005; see also Chan, 2007). John Doyle PLC is a medium-sized group of construction-related companies that started out as a family-run civil engineering firm in 1966 (see **Figure 4.1** for the chronological development of John Doyle PLC). Since the late 1980s, John Doyle PLC went through a series of mergers and acquisitions as part of a deliberate strategy to expand its business offering and increase its professionalisation. As can be seen in **Figure 4.1**, the strategy paid off in terms of increase in the group turnover. John Doyle PLC is now the parent company of four entities, which includes the original civil engineering company (John Doyle Construction Limited), an interior fitting out company (Ibex Interiors Limited), a construction plant-operations and hire company (Blythewood Plant Hire Limited), and a project management company (Bell Projects Limited).

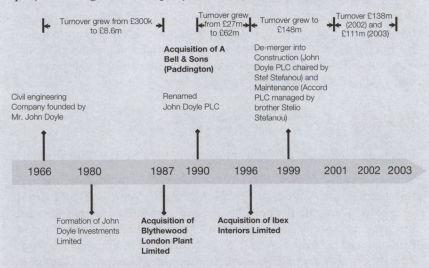

Figure 4.1 Chronological development of John Doyle PLC

People are considered as the most important asset in John Doyle PLC, and this organisational value emanates from the top, from its present Chairman, Mr Stef Stefanou. New managers are constantly being initiated into the group as the Group Secretary explained, 'If a manager does not share this value of respecting the workers, then he will be called in and I will say this is how it is done in John Doyle.' One key performance indicator relating to HRM that is cherished by senior managers in the Group is the labour turnover rate. Across the four companies in John Doyle PLC, labour turnover appeared to be quite low (around 6%). This was attributed to good industrial relations existing as a result of informal interactions between managers and workers, encouraged by the Chairman.

John Doyle PLC operates an 'open-door' policy for all its members of staff to discuss a range of issues, from wage reviews, to discussion of individual performance (e.g. compliance with health and safety policy on construction sites), to career progression opportunities. Despite its informal approach to HRM, there is a conscious attempt to align HRD practices with the organisational strategy of expanding into new markets and increase professionalisation. Developmental opportunities are largely contingent on the individual employee's discretion and initiative. For example, a receptionist employed in John Doyle Construction Limited became interested in working as a quantity surveyor. Through informal conversations with his line manager, it was decided that this was mutually beneficial and the company subsequently sponsored the receptionist on a day-release university degree course. In another instance, a plant operator from Blythewood Plant Hire Limited wanted to expand his knowledge on computers having developed a personal interest in graphic design. He approached his line manager and requested time off to undertake personal study. This coincided with the Group's desire to increase its web presence, and study leave was granted.

Case Study 4.2

Plugging the leakage of human resources at Sunter Plumbing

This case study derives from a research project funded by the UK Engineering and Physical Sciences Research Council (EPSRC) that sought to investigate the links between organisational knowledge and productivity. Specifically, this case study is about a small construction firm, Sunter Limited, operating in the North East of England. The firm directly employs around 60 workers specialising mainly in plumbing, but also in a variety of trades ranging from joinery to wall and floor tilers to heating engineers. Sunter Limited was established in 1973 and traditionally recruited from within the local area where the head office is located. Since its formation, Sunter Limited has developed expertise in the home refurbishment market, and has depended largely on contracts with a single local authority to maintain and upgrade the social housing stock.

With the growing decline in public sector provision of social housing, it became imperative that Sunter Limited did not rely on the local authority client but instead diversify its client base to ensure survival. In order to do this, the directors who founded the company proactively sought to broaden its skills base to permit them to bid for projects beyond the social housing sector. In the late 1990s, the company recruited a manager who possessed extensive experience of working with larger construction companies. As a result, this manager had developed knowledge on contemporary industry trends that could enable the writing of successful bids. One of the first tasks that the manager embarked on was to 'professionalise' the company by widening the appeal of its existing skills to a broader range of prospective clients.

As part of the 'professionalisation' endeavour, the manager considered it critical that employees developed skills beyond technical abilities. So there was a need, recognised by the manager and the directors of the firm, for the workers to be trained in such key skills as communication skills and customer service in order to remain competitive. Furthermore, the manager attempted to formalise HR records so that matters like absenteeism and labour turnover rates could be documented to aid in projecting a professional image when bidding for projects to new clients. As the manager remarked, 'This wasn't easy. The workers had developed the idea that they have always worked for the local authority client, and they could not see why it was necessary to change the way they have done things for the last 30 years.' Thus, a huge part of the manager's early work revolved around persuading the workers of the necessity to enhance their key skills over and above the technical knowledge of their trade. Today, Sunter Limited engages in a wide range of refurbishment projects ranging from schools to doctors' surgeries, and does not depend solely on a single local authority for contracts.

As seen in the cases of John Doyle PLC and Sunter Plumbing on pages 89 and 90, HRD is manifested in a range of activities from the wider sphere of HRM (including recruitment, resourcing, appraisals, and training and development). In both cases, HRD practices were strongly linked to the desire to expand into new markets, as part of a deliberate strategy to grow the business. Where employees had some degree of discretion in identifying personal aspirations for development in John Doyle PLC, a case had to be made to persuade managers that this corresponded with the strategic direction of the business. In the case of Sunter Limited, although the manager was the one convincing employees of the need to professionalise, it is also clear that HRD was intimately linked with the organisational strategy. Arguably, HRD was used as a tactical response to the strategic direction of the firms in both cases, set mainly by its senior managers. At the heart of such alignment, there is a critical performance agenda, measured by indicators such as of labour turnover rates, growth in turnover and profitability, and diversification of markets. Whilst such observations correspond well with the dominant view found in the HRD literature and the linkages between HRD and strategic HRM, the fundamental assumption of alignment is challenged in the next section. Arguably, the notion of alignment serves only to oversimplify the nature of HRD in construction, since the interpersonal, inter-organisational and institutional contexts in which construction operates poses potential problems for the strategic enactment of HRD.

Problematising HRD in construction: exploring the challenges of alignment in project-based construction

In the preceding section, it was suggested that prevailing thoughts about HRD in the literature–as well as in practice – have centred on the need to 'fit' HRD practices with wider organisational strategic concerns. In this section, the notion of 'fit' is questioned as three challenges are raised regarding the alignment between HRD practices and organisational performance and strategy. First, it is suggested that the performative lens through which HRD is viewed focuses too much attention on the agenda of managers and ignores the genuine involvement and participation of workers. The balance has skewed firmly towards a managerial functionalist position of HRD thereby undermining the benefits that HRD practices could accrue to individual workers. Second, the HRD literature has largely taken a blinkered view of organisations as singular, coherent entities, which neglects the inter-organisational relations that are central to the delivery of construction work. The temporary, multi-organisational coalitions that typify the project-based nature of construction means that it is difficult and ambiguous for companies in the construction industry to specify HRD practices in a clear and consistent manner. Thirdly, the institutional context of HRD within which construction firms operate is complex, reflecting both the diversity of the sector and its interaction with the wider political economy of skills formation that in turn influences the choices made by companies to adopt appropriate HRD practices for their market context. This section will consider each of these challenges in turn.

Challenge of recognising managerial and individual agendas of HRD in construction

HRD is normally framed in a way that regards the skills and knowledge of an organisation's human resources as assets to be managed for the sake of improving organisational performance and competitiveness. HRD is often reduced to a managerial process of identifying the strategic orientation of the firm, establishing gaps in competencies and capabilities, and pursuing HRD practices that will plug those gaps. Yet, such a mechanistic view seldom acknowledges the power asymmetries that exist between managers and workers. So, whilst the cases above of John Doyle PLC and Sunter Limited have been interpreted as showing alignment between HRD practices and organisational strategy, a plausible alternative way of looking at those examples would instead highlight the power retained by managers to determine the HRD agenda that places the needs of workers as a secondary concern to the narrowly-defined, unitarist approach of maximising organisational performance.

Academics have long commented on the divergence between the interests of managers and workers in organisational life, and how this influences participation in HRD. When investigating how the productivity agenda is being framed in the UK construction industry, for instance, Chan and Kaka (2007) observed that managers were inclined to place more credence on technical issues of project planning (including the management of human resources) whereas workers emphasised the basics of adequate training and development as a critical issue to sustain a productive workplace. This echoes Druker *et al.'s* (1996) distinctions between 'hard' and 'soft' HRM, as they argued that managers tended to treat human resources as simply another organisational asset to be controlled. Clarke (1992) certainly distinguished the ways in which the notion of skill is conceived by employers and workers: the former treating skills as commoditised assets to be procured and consumed; and the latter regarding skills and their reproduction as a means to preserve a collective identity. Dainty *et al.* (2005) also suggest that employers tend to prevent their workers getting training in the first place to minimise the risk of other employers poaching them after they have completed the training (see also Beckingsdale and Dulaimi, 1997; Forde and MacKenzie, 2007).

Tensions exist within organisations that necessitate the balancing of short-term goals of achieving profitability and the longer-term needs of workers and organisations to sustain the reproduction of skills, competencies and knowledge. Raidén and Dainty (2006) used the phrase 'chaordic organization' to describe just how construction companies deal constantly with both the chaotic business environment and the orderly, strategic development of human resources. Yet, it is maintained that the pendulum often swings too far in favour of employers where HRD is concerned. In both the cases of John Doyle PLC and Sunter Limited, it is clear that whilst employees have been involved in identifying areas for development, the ultimate decision-making power in terms of what to invest in rests with the higher levels of the organisational hierarchy. Employee participation could be seen to be merely symbolic, and not illustrative of the models of co-determination and employee control as articulated by Marchington and Wilkinson (2005). To ensure genuine employee participation in framing the HRD agenda, there is a need for a level playingfield for both managers and workers in organisations. Yet, Chan and Cooper (2006) argue that construction workers are seldom armed with the ability to adequately articulate their HRD needs to their managers because of latent skills shortages (see Keep and Mayhew, 2003). In other words, workers do not reflect on their skills needs because they simply cope with the status quo of poor project resourcing and employee appraisal practices. Furthermore, Grugulis *et al.* (2004) argued that the managerial approach to HRD that stresses the importance of organisational performance serves only to marginalise the autonomy and discretion of workers in negotiating their HRD needs.

The mantra of the knowledge economy is not helpful either for seeking a balance between managerial and employee interests, as Coffield (1999) argues, this tends to be:

> used to socialise workers to the escalating demands of employers, who use: *'empowerment'* to disguise an intensification of workloads via increased delegation; *'employability'* to make the historic retreat from the policy of full employment and periodic unemployment between jobs more acceptable; and *'flexibility'* to cover a variety of strategies to reduce costs which increase job insecurity (p. 488; *original* emphasis).

Indeed, the ability to align HRD practices with organisational strategy and performance remains questionable. Scarbrough (1998) rejects the idea that organisational competencies can be mobilised to satisfy organisational competitiveness, emphasising that 'organisational knowledge is not a biddable resource at the disposal of top management (p. 219).' Marchington and Wilkinson (2008) note the contingency associated with seeking alignment strategies, and remark that a one-size-fits-all approach in the form of 'best practice' is likely to fail. It is often difficult to articulate 'best fit' or 'best practice', since organisational performance depends on an infinite amount of possibilities in configuring bundles of practices, for which HRD is only one constituent part (see Ichniowski *et al.*, 1997; Guest *et al.*, 2004, and; Bloom and Van Reenen, 2010). Even a 'best fit' strategy that is contextually specific to the organisation is problematic. Marchington and Wilkinson (2008) argue that managers do not often have control over their workers, and a top-down, deterministic way of framing strategic options for an organisation does not always work since organisations are rarely coherent and consistent as single entities. Instead, they stress the need to consider the pluralistic and institutional context within which organisations operate, a point that will be explained in the next sub-section.

Challenge of developing skills in multi-employer networked organisations in project-based construction

Organisations are rarely stable, and the process of organising demands the negotiation of divergent interests across a plurality of organisational actors (Vince, 2005). Admittedly, the construction industry has a long-standing tradition of dealing with such a pluralistic view of organisations, since its *modus operandi* is characterised by a diverse range of actors interfacing with one another in temporary project organisations (see Phelps-Brown, 1968; Cherns and Bryant, 1984; and Groák, 1994) that are in constant flux. This creates numerous problems for seeking alignment between HRD practices and organisational strategy *per se*, given the dilemmas faced by construction organisations in developing coherent approaches to HRD and strategic planning. In this sub-section, a number of dilemmas are considered, including problems associated with the diversity of stakeholders found in the construction industry, the 'hollowing-out' of construction organisations that results in the dissolution of the traditional employment relationship, and difficulties in pursuing a learning agenda as a result of ambiguities connected with inter-organisational dynamics in construction.

Dainty and Chan (2011) draw on Pearce (2003) to explain how confusion about which aspects of HRD to pursue arises from the complexity of defining the construction industry and its role. They note that the diversity found in the sector means that the skills needs of construction companies can vary substantially, depending on the degree of specialisation, the size of the firm, and its relative position along the supply chain. This generates the first dilemma in terms of gaining consistency of HRD approaches across the construction industry. The literature on HRD has largely emphasised firm-level analyses that claim aligning HRD practices with the strategic intent of improving business performance and competitiveness is a good thing. However, the empirical basis for such an assertion is often weak (see Coffield, 1999), and the benefits of HRD become distorted when performance and competitiveness at industry sector level is accounted for (see e.g., Clarke and Herrmann, 2004a).

Let us suppose that a major construction company has, as its strategy, to steer towards galvanising its project management expertise. This company then decides to invest in developing its management capabilities to sustain its competitive advantage. As is often the case, the company in question then subcontracts much of its production function to companies within its extensive supply chain. It is possible that benefits might accrue to the major construction company that has strategised to developing its management capabilities as part of moving up the value chain (Porter and Ketels, 2003). However, this could be at the expense of creating skills gaps in other areas of the industry, which could consequently negate any positive effects gained by the single firm. After all, subcontracting has resulted in the proliferation of non-traditional forms of employment, which in turn reduces the propensity of many firms in the industry to engage in HRD activities (see Forde and MacKenzie, 2007). Perhaps, the myopic treatment of HRD at the firm-level can explain the fear of poaching that is endemic across the construction industry. As Clarke and Herrmann (2004b) observe:

> the concern to control costs and contract relations determines all aspects of the construction process and has contributed to a further diminution in production knowledge and skills in British construction (p. 522).

And as they later add

> at the expense of considerations of production and the ways in which
> this determines *all* aspects of the construction process (2004a: 1058,
> *emphasis* added).

Indeed, the diversity of the sector creates paradoxical tensions for firms to
effectively align HRD practices with organisational strategy and performance. The consideration of human capital perspective at the firm-level is
simply inept (Clarke, 2006). Instead, there is a need to consider the HR
architecture and employment sub-systems in construction more finely, and
examine how trade-offs in employment options influence the nature of skills
and knowledge-sharing across groups of workers in the industry (see Lepak
and Snell, 2007).

To complicate matters further, there is also the dissolution of the
traditional employment relationship to contend with. The reliance on self-
employed workers and extensive subcontracting for its production function
has been observed to deter construction employers from investing in HRD.
It is worth noting that where HRD seems to occur in construction companies,
the focus of such activities has centred mainly on professional workers who
now form the core of many companies that specialise in project management
(see Clarke and Herrmann, 2004a; 2004b; and; Raidén and Dainty, 2006).
Indeed, many construction companies have been 'hollowed-out' (see
Castells, 2002), and the production of the built environment can only be
analysed by distilling the inter-organisational dynamics that bind
construction firms across networks (Grugulis *et al.*, 2003; Kinnie *et al.*,
2005; Grimshaw and Rubery, 2005). Yet, networks diminish greatly the
possibility of aligning HRD with strategy and performance, as traditional
firm boundaries become disordered (see Marchington *et al.*, 2009) and
issues regarding who holds the responsibility for, and how firms in
construction actually engage in, HRD become ever more blurred.

Rubery *et al.* (2004) argue that the underlying assumption of the resource-
based view of the firm–that internally consistent HR policies can contribute
to an organisation's strategic objectives–is questionable in reality due to
influences deriving from external sources (see, for example the Case on
Building Futures East on the next page). The ambiguities of control in terms
of HRD activities invokes the individualisation of the employment
relationship where non-standard forms of employment exist (Chan and
McCabe, 2010) and the responsibility for training and learning is gradually
placed on the individual worker (Rubery *et al.*, 2002). The shift towards
individuals taking responsibility for their training and learning needs can
only run counter to exhortations of alignment between HRD practices and
organisational strategy and performance, as self-interests of career
progression prevail over the Utopian view of developing the knowledge-
based organisation characterised by learning and knowledge sharing

(Rubery *et al.*, 2002). Notwithstanding the obstacles of getting individuals to articulate their developmental needs discussed above, the fragmentation of the stable employment relationship and the propagation of networks can also discourage organisational learning in project-based construction (see Chan *et al.*, 2005).

Case Study 4.3

Building coherent strategy or tactical coping? Unlocking local potential in Building Futures East (BFE)

Building Futures East (BFE) is a charity based in the West of Newcastle upon Tyne in England, studied as part of an investigation of institutional perspectives of construction skills development in the UK (see Chan and Moehler, 2008). This locality had a longstanding problem of workless households, and the priest from the local parish church, Fr Michael Conaty, came up with the idea of equipping every unemployed young person in this locality with the skills of a construction trade (e.g. joinery and bricklaying). Fr Conaty enlisted the help of a parishioner who worked for the local authority. At the same time, regeneration efforts were under way near the locality and Fr Conaty was keen to engage with the house-building contractors that had been awarded contracts to build the developments in adjacent localities.

A bid was submitted to the European Social Fund (ESF) to set up a training centre in the deprived locality. This would see the formation of a collaborative network of trainers, local community participants (e.g. workless parents and their youngsters), and housebuilders in the area. This allowed different housebuilders operating near the locality to share the responsibility of developing the skills' capacity of unemployed young people in the area. The engagement of the employers would not have happened otherwise, since taking on the young people would have been perceived to be too risky, and because of the transience of construction work and the construction workforce. With the funding made available through the ESF funding, Fr Conaty was able to realise his vision by setting up BFE to inject skills development, and wider aspirations and employment opportunities in his locality. The efforts paid off, and BFE won numerous awards.

Yet, this endeavour was threatened by the shelf-life of ESF funding and the housebuilding projects, and the financial crisis which caused development activity in this area to sharply decline. To sustain the training activity, and to ensure that the initial efforts to regenerate employment in the locality were not wasted, the trustees of BFE must constantly re-apply for renewed funding . With the formation of the new coalition government in the UK and its agenda of encouraging more social enterprises (i.e. organisations that apply market-based principles for a social purpose), BFE is being rebranded in an attempt to secure successful bids to future funding streams. As part of the rebranding exercise, BFE has commercialised part of its training activities by providing specialist training for car manufacturer, Nissan, in Sunderland. A tension is therefore created for BFE which now has to ensure that it retains its core purpose of training local unemployed young people in a construction trade on the one hand, whilst also restructuring to include more market-based activities.

HRD is often treated as a cost by construction companies and financing this 'burden' can be challenging. Novel ways of organising HRD in a networked fashion (see Gospel and Howard, 2006) can reduce these costs. The transitory nature of construction work means that stabilising the membership of employers within the network can be troublesome. Consequently, BFE remains vulnerable to the external environment (i.e. the financial climate that in turn threatens funding sources) and is simply not a master of its own destiny. What started out as a group of house-building companies successfully collaborating with a local community to develop and enact a coherent skills development strategy for the locality quickly and paradoxically transformed into a difficult situation caused by the completion of that very building work and with it the ending of a funding round. This could be seen to demonstrate the difficulties of seeking any form of alignment between HRD efforts and strategy. New rhetorical forms have had to be mobilised; in this case, with BFE moving from an agenda of local regeneration tosocial enterprise building (with associated challenges of sustaining the original core purpose).

Challenge of negotiating the institutional context of HRD in construction

In the preceding discussion, we have seen how managerial attempts to align HRD practices with organisational strategy and performance are limited by interpersonal and inter-organisational dilemmas. There is also a need to consider the role of the state and how institutional forces can influence decisions made about HRD in organisations. After all, part of the HRD solutions can be found through the VET system coordinated by the state. Indeed, institutional frameworks matter since they are critical in coordinating HRD efforts at an industry level given the complex nature of the concept of skills and the plurality of interests represented across the diverse range of organisations working in the sector. Fundamentally, there should be dialogue coordinated between the three key actors of the state, employers and employees (see Coffield, 1999; Clarke, 2006). These key actors, also known as social partners, have very different interests in shaping strategies for skills development. Clarke and Winch (2007) suggest that the state is largely concerned with enhancing the productive capacity of the society it governs, whereas employers engage in HRD that meets their short-term production needs whilst individuals partaking in lifelong learning are influenced by their ability to enter and/or progress in the labour market. Clarke and Winch argued that:

> The ways in which particular qualities of labour are nurtured, advanced and reproduced [...] tells us a great deal too about the value accorded to labour in society [...] conflicting interests [...] represent a compromise and at the same time reflects the power attached to each of these different interests (p. 1).

Different countries constantly modify the configuration of the interactions between the three parties in different ways. This is in part due to shifts in political ideology and in response to trends such as the globalisation of capital. Nonetheless, whilst there is not a single model for coordinating the divergent interests of the state, employers and individuals, it is evident that each of these stakeholders does feature in many systems of developing vocational skills. For example, Westerhuis (2007) explains how the state in the Netherlands builds, within its VET system, statutory instruments to ensure the engagement of young people in some form of vocational or higher education in a compulsory education system that encourages industry involvement and public engagement. Similarly, Green *et al.* (1999) note the significance of the state in the Asian industrialised economies of Singapore, South Korea and Taiwan in centrally coordinating the education system that aligns employment needs of growing industries with lifelong learning of individuals to facilitate a higher-value-added production economy. Furthermore, Goh and Green (1997) highlighted how the state could embed trade union involvement in a collaborative framework of initiatives that guarantees continuous training and education of workers in Singapore. Notwithstanding the importance of the role of the state, Géhin (2007) recognises the increasingly important role employers play in France to shape the education system: 'the Act of 1993 on labour, employment and vocational training [...] sought to inspire decentralized concerted planning, consultation with financial institutions, unions and employers (2007: 45–6).'

An institutional framework is more than just its coordination; often, the organisation of education and training is also subjected to a philosophical dimension. Most notably, Greinert (2007) discusses the *Beruf* concept as the organising principle in German vocational education; essentially, most Germans do not simply seek out a 'job', but a *Beruf* (a sense of vocation and identity). Greinert (2007) explained how this philosophy has shaped the German VET system through a range of mechanisms, including the explicit framing of the *Beruf* philosophy within the German constitution, the deep engagement of the social partners involving dialogue between trade unions and major employers' associations, and the manifestation of a quasi-standardised route for education, employment and career development. By contrast, vocational education in the US is dominated by the ideology of social efficiency, which in turn leads to the creation of statutory instruments that legitimise vocational education that is job-specific and employer-led (Lewis, 2007). In the UK, the institutional coordination of VET is driven by an employer-led system (see Chan and Dainty, 2007). This is underpinned by a financing system for which employers pay a training levy (currently at 0.5% for direct employees and 1.5% for labour-only-subcontractors if the wage bill is above £80,000 per annum) to, and attempt to claim training grants back from, the sector skills council ConstructionSkills (see www.cskills.org).

The heterogeneity of institutional approaches illustrates how divergent interests in skills development can be coordinated at an industry level, which in turn influences how employers and employees participate in HRD activities. The stronger the institutional frameworks that govern HRD–that is, the maintenance of dialogue and consensus-seeking between the social partners–the better the uptake of HRD activities that in turn contribute to sustained industrial development (see e.g. Broadberry and O'Mahony, 2004; and Leitch, 2006). However, such alignment between the social partners is not unproblematic. Taking the UK for example, social partnership is typically weak, and the lack of industrial democracy is demonstrated by the reduction of employee and trade union involvement in shaping the HRD agenda at an industry level (see Clarke, 2009). To exacerbate matters, Chan and Moehler (2008) observed that the institutional landscape is fraught with panoply of government departments and agencies, private and public sector education and training providers involved in formulating policies, funding and providing for skills development in construction. This in turn confuses employers wishing to up-skill their workforce because there is lack of clarity in terms of the funding and qualifications framework (see e.g. Grugulis, 2003).

Indeed, Chan and Moehler (2008) found that employers who are able to 'manipulate' the institutional system of skills development, for example by utilising appropriate rhetorical language when completing training grant application forms (see Gold and Smith, 2003), can access funding and be seen to engage in skills development. They argued that the weak institutional framework in the UK has dampened the confidence of many small and medium-sized employers to engage with the VET system that should provide much of their HRD needs (see the Case of Gunpoint on the next page). They reported on an employer who would rather pay a premium for its employees to attend the French Compagnonnage training programme known to provide an opportunity for developing high quality craft skills than to engage with the British VET system (see Malloch *et al.*, 2007). Employers who are unable to afford such a privilege revert back to an informal way of learning-by-watching and learning-by-doing for meeting their HRD requirements (Chan *et al.*, 2010).

Case Study 4.4

Making a point for specialist skills in Gunpoint

Gunpoint is a chain of licensed franchisees across the UK specialising in masonry re-pointing work (see www.gunpointlimited.co.uk). The particular franchisee explored here forms part of a broader study by Chan and Moehler (2008), based in the North East of England. Gunpoint North East is a micro-organisation employing less than 50 directly-employed workers. All the workers are qualified to a National Vocational Qualification (NVQ) level 3 in bricklaying. However, because Gunpoint utilises a patented proprietary gun-powered mortar pump and injection system to undertake re-pointing work, there is no existing training provision that can allow the development of a qualified workforce in this area. Yet, there is an urgent need to ensure that the proprietary system is used appropriately and safely!

This is not helped by the institutional framework which the licensee, Fiona Robertson, perceives as creating much confusion for employers in terms of the nature and availability of training provision in this area. Instead, the institutional framework simply maintains training provision in the 'biblical' trades, which are deemed to neglect modern construction methods and technologies (see Chan *et al.*, 2010). At one point, Fiona was frustrated by the response of the sector skills council, ConstructionSkills, who basically asked Gunpoint to 'develop and deliver their inhouse training materials at their own cost.' Not being educators and trainers themselves, Gunpoint was uncertain as to how this could be done, on top of their day-to-day operations. After much negotiation and brokering, ConstructionSkills was able to connect Fiona with a local college to develop appropriate course materials for upskilling her workers. However, it took over 18 months to reach this point, with the workers having to continue working without the necessary requisite skills for the job meantime. And because of that fact – that the workers had coped so well with the task meantime – they saw little value in undertaking the training programme at all, so Fiona had to exert a lot of effort in persuading them to do so.

According to Fiona, the institutional framework could become more responsive to the changing needs of the industry. Employers like Fiona often begrudge not getting a service back from paying the training levy to ConstructionSkills. However, for Chan and Moehler (2008), the system is largely driven by a cost-effective agenda that encourages a great deal of bureaucracy, which in turn promotes only the status quo. Without equitable and effective representation of the social partners in shaping the HRD agenda at an industry level, a weak institutional system will remain inadaptable to changing technological requirements of the industry. A corollary of this is a high dependence on persuasion across the diverse range of stakeholders as a means to frame the HRD system.

Concluding thoughts on implications for research, policy and practice

Generally, the literature has been optimistic about the alignment of HRD with organisational objectives. Benefits of HRD to organisational performance have often been claimed without the necessary foundation of empirical evidence. Yet, the purpose of HRD should be much broader than the narrow margins of the human capital perspective. It has been argued that benefits should also accrue to individual workers, employers and society at large. Indeed, the discussion in this chapter has exposed a number of dilemmas and inconsistencies in the literature that cast doubt on the oversimplified assumption that the aim of HRD is to align with organisational goals of improving performance and enhancing competitiveness. Instead, problems have been identified to show that the espoused benefits of HRD are not always met because of the challenges of integrating the agendas of the corporate and political elite with the needs of employees and organisations across the industry. Furthermore, companies are not always in control of their own destinies, especially in the networked organisational context that has become the norm in construction. The fuzziness of organisational boundaries leads to challenges of controlling HRD responsibilities and determining the link between HRD and organisational strategy. So, what are the implications of these challenges for research, policy and practice?

Research perspective

It is suggested that inter-organisational relations matter considerably in the examination of HRD in construction. Just how the diverse range of stakeholders engage with the HRD agenda, both formally through the institutionalised system and informally in the workplace, remains an under-researched area. The assumption of alignment between HRD and organisational strategy has often acted as a convenient crutch for researchers to simplify the dynamics of HRD and promote a prescriptive view of what HRD should be doing. Yet, as the case studies presented in this chapter illustrate, there is much to be explored in terms of how the powers of negotiation and persuasion are enacted between managers and workers, between different companies across the supply chain, and between employers and the social partners in pursuing skills development options. To fully appreciate the workings of HRD in construction, it is imperative that levels of analysis across interpersonal, inter-organisational and institutional dimensions are clarified.

Policy perspective

One could argue, however, that the dynamics of HRD articulated here is symptomatic of the weak, liberal institutional form illustrated by the UK-centric cases in this chapter. From a policy perspective, therefore, there needs to be a rethink of the arrangement of the social partners in deciding on HRD matters for the industry and economy. Firm-level analyses are inadequate to achieve the aspirations of the lifelong learning agenda and the goals of the knowledge economy. There needs to be equitable partnership that ensures the involvement of a range of employers and employees. This could be done by creating institutionally-coordinated forums that reflect the diversity of the industry. There is a need to learn from countries (e.g. Germany) that have been successful in maintaining equitable social partnership models of institutional coordination to ensure that the focus is firmly placed on enhancing the qualities of skills and knowledge. Unfortunately, given the liberal market philosophy which underpins HRD in the UK, this is unlikely to happen in the near future due to an obsession with short-term financial returns—and cost-cutting—which privileges shareholder returns rather than sustained, long-term performance.

Practice perspective

Finally, from a practice perspective, employers need to develop an ability to reflect on work organisation within and beyond their organisational boundaries, especially in relation to informal arrangements at the work-place. There are tensions at any workplace in terms of balancing the security of benefits to individual workers and the economic objectives of the firm, industry and society. Resolving these tensions is part and parcel of organisational life, but genuine participation of managers and workers in co-determining the HRD agenda needs to be encouraged if the sector is to move beyond perspectives that are driven predominantly by short-term commercial pressures.

References

Armstrong, M. (2003) *A Handbook of Human Resource Management Practice*, 9th edn, London: Kogan Page.

Aronowitz, S. and DiFazio, W. (1994) *The Jobless Future*, Minneapolis: University of Minnesota Press.

Barney, J. (1991) Firm resources and sustained competitive advantage. *Journal of Management*, 17(1), 99–120.

Barney, J. (2007) Looking inside for competitive advantage, R. S. Schuler and Jackson S. E. (eds), *Strategic Human Resource Management*, 2nd edn, Oxon: Blackwell, 9–22.

Becker, G. (1964; 1993) *Human Capital: a theoretical and empirical analysis, with special reference to education*, 3rd edn, Chicago: University of Chicago Press.

Beckingsdale, T. and Dulaimi, M. F. (1997) *The Investors in People standard in UK construction organisations.* CIOB construction papers, 9–13.

Bee, F. and Bee, R. (2003) *Learning Needs Analysis and Evaluation.* London: Chartered Institute of Personnel and Development.

Bell, E., Taylor, S. and Thorpe, R. (2002) A step in the right direction? Investors in People and the learning organisation, *British Journal of Management*, 13, 161–71.

Bloom, N. and Van Reenen, J. (2010) Why do management practices differ across firms and countries? *Journal of Economic Perspectives*, 24(1), 203–24.

Blyton, P. and Turnbull, P. (2004) *The Dynamics of Employee Relations,* 3rd edn, Basingstoke: Palgrave.

Boxall, P., Purcell, J. and Wright, P. (2007) *The Oxford Handbook of Human Resource Management*, Oxford: Oxford University Press.

Broadberry, S. and O'Mahony, M. (2004) Britain's productivity gap with the United States and Europe: a historical perspective, *National Institute Economic Review*, 189, 72–85.

Brockmann, M., Clarke, L. and Winch, C. (2009) Difficulties in recognising vocational skills and qualifications across Europe, *Assessment in Education: principles, policy and practice*, 16(1), 97–109.

Buckley, R. and Caple, J. (2004) *The Theory and Practice of Training*, 5th edn, London: Kogan Page.

Burnham, P. (2001) New Labour and the politics of depoliticisation, *British Journal of Politics and International Relations*, 3(2), 127–49.

Chan, P. and Cooper, R. (2006) Talent management in construction project organisations: do you know where your experts are?, *Construction Information Quarterly*, 8(1), 12–18.

Chan, P. and Kaka, A. (2007) The impacts of workforce integration on productivity, A. Dainty, S. Green and B. Bagilhole (eds), *People and Culture in Construction: a Reader.* London: Spon, 240–57.

Chan, P. and Moehler, R. (2008) Construction skills development in the UK: transitioning between the formal and informal, in *Proceedings of the CIB W55/W65 joint conference: transforming through construction*, 15–17 November 2008, Dubai, UAE.

Chan, P. W. and McCabe, S. (2010) Emerging disparities: Exploring the impacts of the financial crisis on the UK construction labour market, C. O. Egbu and Lou E. W. (eds), *Proceedings of the 26th ARCOM conference*, 6–8 September 2010, Leeds, UK, Association of Researchers in Construction Management, 523–32.

Chan, P. W. and Räisänen, C. (2009) Informality and emergence in construction, *Construction Management and Economics*, 27(10), 907–12.

Chan, P. W., Cooper, R. and Tzortzopoulos, P. (2005) Organisational learning: conceptual challenges from a project perspective, *Construction Management and Economics*, 23(7), 747–56.

Chan, P., Clarke, L. and Dainty, A. (2010) The dynamics of migrant employment in construction: can supply of skilled labour ever match demand?, M. Ruhs and Anderson, B. (eds), *Who Needs Migrant Workers? Labour shortages, immigration, and public policy.* Oxford: Oxford University Press, 225–55.

Cherns, A. B. and Bryant, D. T. (1984) Studying the client's role in construction, *Construction Management and Economics*, 2, 177–84.

Clarke, L. (1992) *The building labour process: problems of skills, training and employment in the British construction industry in the 1980s*, Occasional Paper No. 50, CIOB, Englemere.

Clarke, L. (1999) The changing structure and significance of apprenticeship with special reference to construction, P. Ainley and Rainbird, H. (eds), *Apprenticeship: towards a new paradigm of learning*, London: Kogan Page, 25–40.

Clarke, L. (2006) Valuing labour, *Building Research and Information*, 34(3), 246–56.

Clarke, L. (2007) The emergence and reinforcement of class and gender divisions through vocational education in England, in Clarke, L. and Winch, C. (2007) (eds), *Vocational education: international approaches, developments and systems*, Abingdon: Routledge, 62–76.

Clarke, L. (2009) Introduction: employee and trade union involvement in VET, J. Cremers (ed), *CLR News 1/2009*, Brussels: European Institute of Construction Labour Research (CLR).

Clarke, L. and Herrmann, G. (2004a) Cost vs production: labour deployment and productivity in social housing construction in England, Scotland, Denmark and Germany, *Construction Management and Economics*, 22(10), 1057–66.

Clarke, L. and Herrmann, G. (2004b) Cost vs production: disparities in social housing construction in Britain and Germany, *Construction Management and Economics*, 22(5), 521–32.

Clarke, L. and Herrmann, G. (2007) Divergent divisions of construction labour, A. Dainty, Green S. and Bagilhole B. (eds), *People and Culture in Construction: a Reader*. Abingdon: Spon, 85–105.

Clarke, L. and Winch, C. (2004) Apprenticeship and applied theoretical knowledge. *Educational Philosophy and Theory*, 36(5), 509–21.

Clarke, L. and Winch, C. (2006) A European skills framework? But what are skills? Anglo-Saxon versus German concepts, *Journal of Education and Work*, 19(3), 255–69.

Clarke, L. and Winch, C. (2007) (eds), *Vocational Education: international approaches, developments and systems*, Abingdon: Routledge.

Coffield, F. (1999) Breaking the consensus: lifelong learning as social control, *British Educational Research Journal*, 25(4), 479–99.

Commission of the European Communities (CEC), (2000) *A memorandum on lifelong learning*, Commission Staff Working Paper (SEC (2000) 1832, 30.10.00), Brussels: CEC.

Commission of the European Communities (CEC), (2001) *Communication from the Commission: making a European area of lifelong learning a reality* (COM (2001) 678, final, 21.11.01), Luxembourg: Office for Official Publications of the European Communities.

Cooper, R. and Chan, P. (2005) *Mapping of skills and inter-organisational competencies (MoSaIC): Case study of John Doyle PLC*, November. Salford: Salford Centre of Research and Innovation in the Built and Human Environment (SCRI).

Dainty, A. and Chan, P. (2011) Human resource development: rhetoric, reality and opportunities, P. Chinowsky and Songer A. (eds), *Organization Management in Construction*, London: Taylor & Francis, 76–94.

Dainty, A. R. J., Ison, S. G. and Root, D. S. (2005) Averting the construction skills crisis: a regional approach, *Local Economy*, 20(1), 79–89.

DiMaggio, P. J. and Powell, W. W. (1983) The iron cage revisited: institutional isomorphism and collective rationality in organizational fields, *American Sociological Review*, 48(2), 147–60.

Druker, J. and White, G. (1996) *Managing People in Construction,* London: Institute of Personnel and Development.

Druker, J., White, G., Hegewisch, A. and Mayne, L. (1996) Between hard and soft HRM: human resource management in the construction industry, *Construction Management and Economics*, 14, 405–16.

Garavan, T. N., Heraty, N. and Barnicle, B. (1999) Human resource development literature: current issues, priorities and dilemmas, Journal of European Industrial Training, 23(4), 169–79.

Géhin, J-P (2007) Vocational education in France: a turbulent history and peripheral role, Clarke, L. and Winch, C. (2007) (eds), *Vocational education: international approaches, developments and systems*, Abingdon: Routledge, 34–48.

Goh, E. and Green, F. (1997) Trade unions as agents for skill formation: the case of Singapore, *International Journal of Training and Development*, 1(4), 230–41.

Gold, J. and Smith, V. (2003) Advances towards a learning movement: translations at work. *Human Resource Development International*, 6(2), 139–52.

Gospel, H. and Foreman, J. (2006) Inter-firm training coordination in Britain, *British Journal of Industrial Relations*, 44(2), 191–214.

Grant, R. (1996) Towards a knowledge-based theory of the firm, *Strategic Management Journal*, 17, 109–22.

Green, F., Ashton, D., James, D. and Sung, J. (1999) The role of the state in skill formation: evidence from the Republic of Korea, Singapore and Taiwan, *Oxford Review of Economic Policy*, 15(1), 82–96.

Greinert, W-D (2007) The German philosophy of vocational education, Clarke, L. and Winch, C. (2007) (eds), *Vocational Education: international approaches, developments and systems*, Abingdon: Routledge, 49–61.

Grimshaw, D. and Rubery, J. (2005) Inter-capital relations and the network organisation: redefining the work and employment nexus, *Cambridge Journal of Economics*, 29(6), 1027–51.

Groák, S. (1994) Is construction an industry? Notes towards a greater analytic emphasis on external linkages, *Construction Management and Economics*, 12, 287–93.

Grugulis, I. (2003) The contribution of National Vocational Qualifications to the growth of skills in the UK, *British Journal of Industrial Relations*, 41(3), 457–75.

Grugulis, I. (2007) *Skills, Training and Human Resource Development: a critical text*, Basingstoke: Palgrave Macmillan.

Grugulis, I. and Bevitt, S. (2002) The impact of Investors in People: a case study of a hospital trust, *Human Resource Management Journal*, 12(3), 44–60.

Grugulis, I., Vincent, S. and Hebson, G. (2003) The rise of the 'network organisation' and the decline of discretion, *Human Resource Management Journal*, 13(2), 45–59.

Grugulis, I., Warhurst, C. and Keep, E. (2004) What's happening to 'skill'?, Warhurst, C., Keep, E. and Grugulis, I. (eds), *The Skills that Matter*, Hampshire: Palgrave Macmillan, 1–19.

Guest, D., Conway, N. and Dewe, P. (2004) Using sequential tree analysis to search for 'bundles' of HR practices, *Human Resource Management Journal*, 14(1), 79–96.

Harrison, R. (1997) *Employee Development,* London: Institute of Personnel and Development.

Holton, E. F. III (1996) The flawed four-level evaluation model, *Human Resource Development Quarterly,* 7(1), 5–21.

Holton, E. F., III (205) Holton's evaluation model: new evidence and construct elaborations, *Advances in Developing Human Resources,* 7(1), 37–54.

Holton, E. F., III, Bates, R. A. and Ruona, W. E. A. (2000) Development of a generalized learning transfer system inventory, *Human Resource Development Quarterly,* 11(4), 333–60.

Huselid, M. A. (1995) The impact of human resource management practices on turnover, productivity, and corporate financial performance, *Academy of Management Journal,* 38(3), 635–72.

Ichniowski, C., Shaw, K. and Prennushi, G. (1997) The effects of human resource management practices on productivity: a study of steel finishing lines, *The American Economic Review,* 87(3), 291–313.

Keenoy, T. (1999) HRM as hologram: a polemic, Journal of Management Studies, 36(1), 1–23.

Keep, E. (2007) The multiple paradoxes of state power in the English education and training system, Clarke, L. and Winch, C. (2007) (eds), *Vocational Education: international approaches, developments and systems,* Abingdon: Routledge, 161–75.

Kinnie, N. J., Swart, J. and Purcell, J. (2005) Influences on the choice of HR system: the network organisation perspective, *International Journal of Human Resource Management,* 16(6), 1004–28.

Kululanga, G. K., Edum-Fotwe, F. T. and McCaffer, R. (2001) Measuring construction contractors' organisational learning, *Building Research and Information,* 29(1), 21–9.

Langford, D. A. and Hughes, W. P. (eds), (2009) *Building a Discipline: the story of construction management,* Reading: Association of Researchers in Construction Management (Arcom).

Lee, M. (2001) A refusal to define HRD, *Human Resource Development International,* 4(3), 327–41.

Lee, M. (2003) The complex roots of HRD, in M. Lee (ed), *HRD in a Complex World,* London: Routledge, 7–24.

Legge, L. (2005) *Human Resource Management: rhetorics and realities.* Basingstoke: Palgrave Macmillan.

Leitch, A. (2006) *Prosperity for All in the Global Economy: work class skills,* final report, December. Norwich: HMSO.

Lepak, D. and Snell, S. A. (2007) Employment subsystems and the 'HR architecture', P., Boxall, J., Purcell P. Wright (eds), *The Oxford Handbook of Human Resource Management,* Oxford: Oxford University Press, 210–30.

Littler, C. R. and Innes, P. (2003) Downsizing and deknowledging the firm, *Work, Employment and Society,* 17(1), 73–100.

Loosemore, M., Dainty, A. and Lingard, H. (2003) *Human Resource Management in Construction Projects: strategic and operational approaches,* London: Spon.

McCracken, M. and Wallace, M. (2000) Exploring strategic maturing in HRD: rhetoric, aspiration or reality, *Journal of European Training,* 21(8/9), 425–46.

McGoldrick, J., Stewart, J. and Watson, S. (2001) Theorizing human resource development, *Human Resource Development International,* 4(3), 343–56.

Malloch, H., Kleymann, B., Angot, J. and Redman, T. (2007) *Les Compagnons du Devoir*: a French Compagnonnage as a HRD system, *Personnel Review*, 36(4), 603–22.

Marchington, M. and Vincent, S. (2004) Analysing the influence of institutional, organizational and interpersonal forces in shaping inter-organizational relations, *Journal of Management Studies*, 41(6), 1029–56.

Marchington, M. and Wilkinson, A. (2005) Direct participation, S. Bach (ed), *Managing Human Resources: Personnel management in transition*, 4th edn, Oxford: Blackwell, 398–423.

Marchington, M. and Wilkinson, A. (2008) *Human Resource Management at Work: people management and development,* 4th edn, London: Chartered Institute of Personnel and Development.

Marchington, M., Carroll, M., Grimshaw, D., Pass, S. and Rubery, J. (2009) *Managing People in Networked Organisations,* London: Chartered Institute of Personnel and Development.

Nyhan, B., Cressey, P., Tomassini, M., Kelleher, M. and Poell, R. (2004) European perspectives on the learning organisation, *Journal of European Industrial Training*, 28(1), 67–92.

O'Donnell, D., McGuire, D. And Cross, C. (2006) Critically challenging some assumptions in HRD, *International Journal of Training and Development*, 10(1), 4–16.

Orlitzky, M. and Frenkel, S. J. (2005) Alternative pathways to high performance workplaces, *International Journal of Human Resource Management*, 16(8), 1325–48.

Paauwe, J. (2004) *HRM and Performance: achieving long-term viability,* Oxford: Oxford University Press.

Pearce, D. (2003) *The Social and Economic Value of Construction: the construction industry's contribution to sustainable development*, London: nCRISP.

Pfeffer, J. (1998) *The Human Equation: Building profits by putting people first,* Boston, MA: Harvard Business School Press.

Phelps-Brown, E H (1968) *Report of the Committee of Inquiry under Professor E H Phelps Brown into certain matters concerning labour in building and civil engineering*, London: HMSO.

Porter, M. and Ketels, C. (2003) *UK Competitiveness: Moving to the Next Stage*, London: Department of Trade and Industry.

Raidén, A. B. and Dainty, A. R. J. (2006) Human resource development in construction organisations: an example of a 'chaordic' learning organisation?, *The Learning Organisation*, 13(1), 63–79.

Rubery, J., Carroll, M., Cooke, F. L., Grugulis, I. and Earnshaw, J. (2004) Human resource management and the permeable organisation: the case of the multi-client call centre, *Journal of Management Studies*, 41(7), 1199–222.

Rubery, J., Earnshaw, J., Marchington, M., Cooke, F. L. and Vincent, S. (2002) Changing organisational forms and the employment relationship, *Journal of Management Studies*, 39(5), 645–72.

Sambrook, S. (2007) Exploring the notion of 'time' and 'critical' HRD, J. Stewart, Rigg C. and Trehan K. (eds), *Critical Human Resource Development: beyond orthodoxy*, Harlow: Pearson, 23–42.

Sambrook, S. (2009) Critical HRD: a concept analysis, *Personnel Review*, 38(1), 61–73.

Scarbrough, H. (1998) Path(ological) dependency? Core competencies from an organisational perspective, *British Journal of Management*, 9(3), 219–32.

Senge, P. M. (1990) *The Fifth Discipline: the art and practice of learning organisation,* New York: Doubleday Dell.

Sisson, K. and Storey, J. (2000) *The Realities of Human Resource Management: managing the employment relationship,* Milton Keynes: Open University Press.

Smith, S. M. (2009) Enhancing performance with Investors in People recognition: exploring the alleged link *Development and Learning in Organizations*, 23(2), 16–18.

Stasz, C. (2001) Assessing skills for work: two perspectives, *Oxford Economic Papers*, 3, 385–405.

Stata, R. (1989) Organisational learning: the key to management innovation, *Sloan Management Review*, Spring, 63–74.

Swanson, R. A. (2007) Theory framework for applied disciplines: boundaries, contributing, core, useful, novel, and irrelevant components, *Human Resource Development Review*, 6(3), 321–39.

Swanson, R. A. and Holton, E. F., III (2009) *Foundations of Human Resource Development*, 2nd edn, San Francisco: Berrett-Koehler.

Swart, J., Mann, C., Brown, S. and Price, A. (2005) *Human Resource Development: strategy and tactics*, Oxford: Elsevier Butterworth-Heinemann.

Thursfield, D. (2001) Employees' perceptions of skill and some implications for training in three UK manufacturing firms. *Human Resource Development International*, 4(4), 503–19.

Truss, C., Gratton, L., Hope-Hailey, V., McGovern, P. and Stiles, P. (1997) Soft and hard models of human resource Management: a reappraisal, *Journal of Management Studies*, 34(1), 53–73.

Valentin, C. (2006) Research human resource development: emergence of a critical approach to HRD enquiry, *International Journal of Training and Development*, 10(1), 17–29.

Vince, R. (2005) Ideas for critical practitioners, in C. Elliott and Turnbull S. (eds), *Critical Thinking in Human Resource Development,* Abingdon: Routledge, 26–36.

Wernerfelt, B. (1984) A resource-based view of the firm, *Strategic Management Journal*, 5(2), 171–80.

Westerhuis, A. (2007) The role of the state in vocational education: a political analysis of the history of vocational education in the Netherlands, Clarke, L. and Winch, C. (eds) (2007) *Vocational Education: international approaches, developments and systems*, Abingdon: Routledge, 21–33.

Winterton, J. (2007) Training, development and competence, P. Boxall, Purcell, J., Wright, P. (eds), *The Oxford Handbook of Human Resource Management.* Oxford: Oxford University Press, 324–43.

Wood, S. (1999) Human resource management and performance, *International Journal of Management Reviews*, 4(1), 367–413.

Wood, S. and DeMenezes, L. (1998) High commitment management in the UK: Evidence from the workplace industrial relations survey, and employers' manpower and skills practices survey, *Human Relations*, 51, 485–515.

Woodall, J. (2003) HRD: An indisputably virtuous endeavour?, *Human Resource Development International*, 6(2), 137–38.

Wright, P. M., Dunford, B. B. and Snell, S. A. (2007) Human resources and the resource-based view of the firm, R. S. Schuler and S. E. Jackson (eds), *Strategic Human Resource Management*, 2nd edn, Oxon: Blackwell, 76–97.

Youndt, M., Snell, S. and Lepak, D. (1996) Human resource management, manufacturing strategy and firm performance, *Academy of Management Journal*, 39(4), 836–66.

5 Competing on identity rather than price: a new perspective on the value of HR in corporate strategy and responsibility

Martin Loosemore, Dave Higgon and Don Aroney

Introduction

The relationship between the management of human resources and competitive advantage should be placed in the context of fundamental changes over the last two decades, in the way that construction organisations are structured and employees managed. These changes have broadly reflected wider cross-sector trends away from centralised, beaurocratic and autonomous forms of organisation towards decentralised market-driven structures driven by a desire to off-load risk and make organisations more responsive, customer-focussed, flexible, efficient and competitive (Salaman and Asch 2003). One consequence of this trend is that many construction firms have reorganised into independent entrepreneurial silos which serve different markets and customers. Many large firms have also organised themselves into separate functional departments covering the entire procurement process from inception to operation (e.g. development, finance, design, estimating, planning, construction, facilities management etc). While relatively easy to manage into cost centres with clearer accountability to overall business performance, these fragmented structures raise numerous challenges. Among other things, one result can be internal departments failing to communicate effectively, and making independent decisions outside the wider interests of the firm. Conflicting subcultures, systems and procedures can also develop within these firms and responsibility for risk management can often become obscured and confused. In this type of structure, internal management controls are often replaced by market controls meaning that firms become a collection of semi autonomous units surviving in relative isolation. From a human resource management perspective, this type of market-based structure is typically morally justified by arguing that it releases employees from slavish adherence to rigid and centralised corporate rules and procedures giving them control over their lives by enabling them to act autonomously and creatively (Peters and Waterman 1982). However, Green (2009) and Murray and Dainty (2009) argue that central to the successful application of such an approach is the transformation of the employee and the redefinition of their relationship with an organisation in ways which often have profound and negative implications for both employees and firms. While flexible organisations without rigid rules may seem attractive on the surface, the boundaries of these firms and their employee relationships and responsibilities become blurred, long-term social and psychological contracts are replaced by short-term casualised economic contracts and organisational competencies, culture, identity, trust and employee loyalty is diluted. Apart from the many moral questions such forms of organisation raise, they are also recognised to have numerous potential performance implications in terms of reduced safety, product and service quality and time and cost performance (NEDO 1983, Kumaraswamy 1996, Lingard and Holmes 2001).

Ambitious national reform programs have been adopted in many countries to address these moral and performance challenges. However, Ness (2010) argues that a market philosophy still drives management practice in the construction industry, despite considerable rhetoric to the contrary. Underlying Ness' arguments is an ethic which contends that this trend has subjugated human resources to a dispensable resource which is primarily used as a means to a commercial end and which is seen as a liability rather than an asset. This is supported by Loosemore and Phua's (2010) research which found that while many firms in the construction industry are making overtures and efforts to be more socially responsible, many remain driven by 'the bottom line'. However, they also point out that this is not surprising and is likely to remain the case given that construction organisations are not social enterprises. Loosemore and Phua found that ideological and emotionally based arguments about corporate social responsibility, no matter how passionately and convincingly put are unlikely to hold sway with boards of directors unless they are supported by a clear business case. Accepting this, they concluded that the interests of human resources may be better advanced by recognising this reality and working within it, by connecting human resource management to competitive advantage, rather than taking a *purely* moral view: a point of departure from the emerging anti-performative agenda in construction management research. The emerging anti-performative agenda has undoubted value in promoting the important ethical and moral dimensions of management decision-making in the construction sector. However, given Loosemore and Phua's research, to suggest that this be exclusive of economic considerations is both unrealistic, unachievable and possibly counter-productive. Recognising that the relationship between corporate social responsibility, human resource management and organisational performance is contested and problematic, this chapter seeks to explore these relationships in a new way – by triangulating the notion of organisational identity with human resource management and corporate strategy. As outlined later in our case study, we explore how one organisation which has operated in both the centralised bureaucratic and 'siloed' modes, has managed to use its corporate identity to balance these competing demands.

Traditional approaches to strategy

Traditionally, strategic management is the process by which an organisation identifies how it will achieve competitive advantage by defining the activities it will be involved in, the areas where it will and will not compete, and the way in which it will compete in those chosen areas (Porter 1980, Aaker 1998). Traditionally, this involves identifying, choosing and implementing activities, raising and allocating resources, setting priorities, directing, communicating, leading and creating ongoing compatibility between the internal skills and resources of an organisation and the changing business environment in which it operates (Chandler 1962, Ansoff 1965, Porter 1980, 1985a 1985b, Viljoen and Dann 2000). Strategic management is the primary means by which managers maintain competitive advantage, and deliver sustainable business practices, in the context of many changes that take place within and outside an organisation's political, legal, regulatory, cultural, social, technological, economic and ecological environments. An organisation exists so far as it produces a service or product that society wants and thus it must be responsive to what society wants and how society wants it produced and delivered. An organisation also takes its resources (human, physical, financial etc) from its environment and must respond to the nature and availability of those resources and the constraints and costs which are imposed on their use. For example, changes in consumer tastes for greener buildings, advances in information technology, the recent economic crisis, climate change and carbon trading, demographic changes such as the aging workforce, increased migration and skills shortages etc. will fundamentally impact upon future strategy in the construction sector. This will change the way in which firms are structured and governed, how resources such as people, finance and technology are acquired and managed and how products and services are produced and delivered to market.

So traditionally, strategy has been about positioning a firm, in advance, to take maximum advantage of such changes and moving it from a current position to a desired future position, pinpointed in response to clearly defined goals and predicted future trends in the business environment. Strategy is thus a constantly evolving process and is an activity which is orientated towards the long-term not the short-term, although the time-span of any vision does vary depending on the nature of the organisation. For example, a small construction firm or sub contractor involved in small-scale routine domestic projects is likely to have a shorter strategic horizon than a large multinational involved in innovative multimillion dollar commercial projects. The time-span of strategy is also assumed to be related to the level at which it is developed. For example, traditionally, strategy occurs at three main levels, ideally in an integrated and coordinated approach, namely: *corporate, business* and *operational* (Gooderham 1998). Corporate strategy is developed by the board of directors and is concerned with longer term trends. It deals with the nature of the business that the organisation is in (the types of activities that it concerns itself with). Business strategy is developed by senior managers and is concerned with how the organisation operates and competes in each chosen type of activity to achieve strategic goals. Finally, operational strategy is concerned with how business strategy is implemented to achieve business goals. This concept of strategic 'life-span' has been used at Brookfield Multiplex (see Case Study 5.1 on p. 124 of this Chapter) but adapted to recognise the contribution that all organisational levels can make to business strategy. Brookfield Multiplex aligns its approach to strategy with the seniority of its management's positions, as illustrated in **Fig. 5.1**. Focus and responsibility for strategic planning and outcomes is held at the level which has the greatest capacity, and is best suited to deal with it. However, although respective management teams focus on their given areas of responsibility, they can also play a role in both influencing and delivering the broader strategy.

Strategy is also traditionally focused on improving organisational *effectiveness* (long-term fit with business environment and what it does) as distinct from *efficiency* (with how things are done and how it operationalises what it does) (Porter 1996). While fine in theory, this is problematic for many construction firms because performance management systems in construction tend to reward efficiency rather than effectiveness. As Loosemore and Phua (2010) argue, the construction sector seems to be preoccupied with efficiency over effectiveness, leading to an endless pursuit

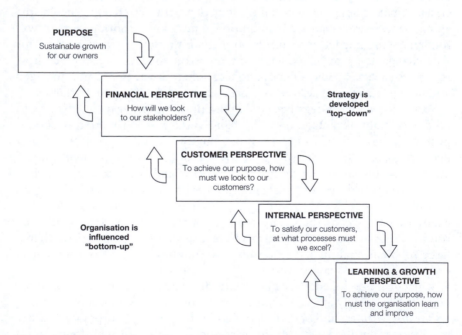

Figure 5.1 Strategic roles and responsibilities and areas of focus in Brookfield Multiplex (see Case Study 5.1).

of lower cost as the primary source of competitive advantage rather than of innovating to develop newer, distinctive and more effective solutions. Inevitably this reduces industry performance to the lowest common denominator; advantages those firms who do not invest in human resource development; and reduces margins to unsustainable levels. Ironically, this also erodes the appetite for larger organisations to invest in innovative pursuits which have often been the primary drivers of efficiency in the first place.

Finally, in addition to achieving effectiveness, strategy is also traditionally concerned with creating operational flexibility (an ability to respond and adapt to the unexpected; Starr *et al.* 2003). It is assumed that strategic errors caused by not understanding or misinterpreting trends in the business environment can be devastating for any organisation and where volatility is high, companies must invest in resources which are adaptable and flexible. Traditionally, in terms of HRM, this has involved strategies such as casualisation, outsourcing, multi-skilling, job sharing and more decentralised management structures (Quinn and Voyer 1996). In construction, many of these strategies have been influenced by the emergence of many 'best practice' management techniques such as lean, outsourcing, de-layering, total quality management, business process re-engineering and benchmarking etc (Green 2006, 2009). Like Pascale (1991), Green has argued that despite the rhetoric of providing flexibility and resilience, in reality these so-called 'best practice' solutions represent stumbling blocks to longer-term and sustainable organisational success by institutionalising rigidity, dehumanising management and eliminating the innovative capacity that is critical to success. This irony appears to be lost on those leaders of industry that so enthusiastically push such reforms.

Questioning traditional notions of strategy from a HR perspective

The traditional systems-based notion of strategy as described above conceptualises strategy as a high-level decision-making process designed to 'fit' a firm to its external environment. Strategy development is seen as a rational, top-down planning function detached from the minutia of day-to-day tasks and organisations are seen as a triangle, with strategists at the top with primarily downward flows of directions, instructions, plans and information to the bottom. While still very influential in shaping the way that strategy is developed in the construction industry (Langford and Male 2001, Cheah and Garvin 2004, Green *et al.* 2008) this view of strategy has been widely criticised. For example, the assumption that strategy can be developed in a rational, systematic and objective way, that managers collect and analyse data with detachment and then review all the options and make a rational decision based on the apparent facts is clearly *not* what happens in practice (Whittington 1993). Furthermore, traditional models of strategy assume a certain sequence of events which is fine in theory but in reality is both limiting and inaccurate. This logical step-by-step process starts with an environmental scan to understand the business environment and then moves systematically to the development and implementation of strategies to adjust and position the organisation to exploit market opportunities (Pettigrew *et al*. 2006). As Hillebrandt and Cannon (1990) and Green *et al*.'s (2008) research into strategy development in construction firms shows, strategy in reality, is more often emergent than pre-planned, shaped by unexpected opportunities and by individuals and often maverick behavior, rather than in response to any formal mechanisms and processes. They found that while boards may intervene with planned strategies occasionally, there was little evidence that formal strategic planning techniques were used in construction firms or that they had much impact on enacted strategy.

Other criticisms of the traditional top-down view of strategy relate to its inability to recognise and integrate the role of legitimate stakeholders in strategy development and its rigidity in responding to uncertain business environments (Hubbard *et al*. 2002). Furthermore, according to Cummings (2005) it has configured the field of strategy to under-value the views and perspectives of people in lower reaches of the organisation, outside stakeholders and communities in favour of a focus on objectivity, economic goals, technology, function and measurable efficiency. Cummings argues that despite what traditional strategy theory argues, the interactions that are crucial to strategy do not occur between executives and the environment at the top of an organisation but between operatives at the coal face who respond quickly to customer needs and who can quickly spot market gaps and opportunities. In reality, the information from these lower level interactions gradually filter-up an organisation to be formally enshrined in high level strategies and plans. Cummings also argues that at the very forefront of strategy as a field of practice lies the recognition that in an increasingly connected and homogeneous business environment where most firms have similar cost structures, supply chains, prices, technologies and systems etc, it is increasingly difficult for firms to differentiate between their business strategies and compete on anything other than margins (a point that resonates within the construction sector). Nevertheless, while these factors are no longer the source of competitive advantage they once were, many firms continue to compete on the same basis as they always have. What distinguishes the most successful firms is a move away from this outdated strategy of competing on price (or more precisely margins) to one which focuses on the one thing that can't be readily copied – their unique corporate identity (bound up in its history, traditions, heritage, core competencies and human resources). Competing on one's unique identity is a completely different approach to strategy – which involves trying to understand one's unique 'ethos' and using this to inspire competitive differentiation. It involves seeing a firm as having a unique individual character, a personality, a credo, a distinctive spirit and a set of differentiating virtues which are rooted in its past stories, personalities, experiences and people. Corporate identity reflects how the organisation 'looks' to its internal and external stakeholders and the most successful companies use this to inspire a distinct style or 'strangeness' in how they want to compete for business. The more a firm devises strategy without leveraging its own unique identity (which is largely based around its people, its culture, its stories, its history and its traditions) then the more like its competitors it will become and the more likely it will be forced to continue down the lowest price (and margin) path. The model which has been utilised effectively at Brookfield Multiplex echoes this approach whereby the organisation has recently gone through an extensive process of consulting its internal and external stakeholders to identify its unique 'DNA', and how it should embed this into its strategic planning, the development of its staff and its operations as a whole. Utilising a strategic framework which adopts similar principles outlined by Cummings, strategy is influenced bottom up, yet driven top down as illustrated in **Fig. 5.2**.

Most recently, profound changes in the business environment with the expansion of information and communication technologies and the increasing connectivity of the global business environment are catalysing new ways to think about strategy and indeed the definition and role of human resources in achieving competitive advantage. Many companies are increasingly focusing on strategic business approaches which look beyond the internal resources of individual business units to take strategic alliances with external partners, mergers and acquisitions and strategic outsourcing etc. (Kodama 2007). Increasingly, sources of creativity, innovation and competitive advantage in business are not being internally generated but are emerging from the confluence of knowledge, experience and ideas from

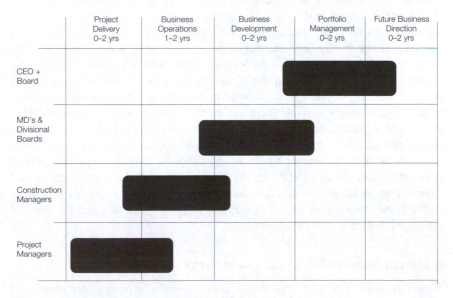

Figure 5.2 Strategy development at Brookfield Multiplex.

many different disciplines, areas of the business and from external sources such as customers, business partners, suppliers, subcontractors, universities etc. The idea that knowledge is co-created in partnerships and communities of practice and creation leads to new business models that synthesise knowledge from a wide area in an open networked system to drive and enhance competitive advantage. It also causes us to look differently at the boundaries of our HRM practices and our traditional confined definitions of what constitute our human resources and our responsibilities towards them. Furthermore, it causes us to question the efficacy of vertically hierarchical and integrated business models (which have a strong tendency to create internal boundaries between disciplines, functions, units, departments and regions, in favour of horizontally dispersed network models with permeable boundaries). This idea of strategic communities of practice also introduces new governance challenges of managing such dispersed structures.

As Stewart (2009) points out, the conventional view of management and strategy is that it is a kind of technology – a technical discipline underpinned by a bundle of techniques which have been scientifically proven. But the contemporary view is that it is a synthetic, creative and entrepreneurial activity bound up in corporate culture and identity rather than a ritualistic, reductionist, risk averse and analytical one which is easily reducible to a number of simple steps and procedures. Strategy he argues, in the way it is traditionally understood and taught, grossly undervalues the role of labour in wealth maximisation and competitive advantage, lacks scientific and scholarly merit and a sound theoretical base and is not yet well understood. This is a technocratic view which he argues legitimises and advances the interests of managers as the ruling managerial class, represents an assault on the power of labour and its importance to our economic prosperity and has led managers seriously astray in seeking pretended technological solutions to what are essentially human, moral, social and political problems. Written before the global financial crisis, Stewart's analysis positions strategy as a neglected branch of the humanities rather than an illusory science and is prophetic in arguing that it is a view of management which causes managers to see organisations serving them rather than the other way around. This approach to management and strategy, he argues, achieves its greatest expression in the '*shareholder-value maximisation model*' of corporate governance that dominates the business schools. It is an approach which Loosemore and Phua (2010) argue, also dominates the construction and engineering sector and which leads most firms to engage in highly repetitive and traditional approaches to strategy which provide very little source of differentiated competitive advantage.

A human resource-based view of strategy

The criticisms above, of traditional approaches to strategy, are clearly of great relevance to this book's objective of better integrating strategy and human resources in the construction sector. To this end it is worth exploring a different but well-established model of strategy which has evolved in response to these concerns – '*the resource-based view*' (RBV). The RBV of strategy links available resources to firm capabilities and competitive advantage by arguing that a firm is made up of a unique pool of resources, namely: physical (land, premises, plant etc); financial (cash, capital etc); human (people, experience and expertise etc) and; organisational (culture, reputation, relations etc) which are greater than the sum of the parts and which interact in unique ways in a web of internal and external environmental relationships (Prahalad and Hamel 1990, Barney 2001). Competitive advantage comes from not simply what resources an organisation owns but how an organisation uses, develops and combines them to produce unique internal and dynamic capabilities which provide competitive advantage (Grant 1996). Competitive advantage also comes from knowing and developing the resources which are a source of a firm's core capabilities that provide competitive advantage. Continual development of a firm's resources is critical because their relevance and uniqueness diminish over time through imitation by competitors and advances in science and knowledge (Teece *et al.* 1997).

This alternative view of strategy development is simply illustrated in **Fig. 5.3**.

In contrast to traditional models of strategy, which argue that organisations should be designed around strategies, the RBV perspective argues that strategies should be built around unique organisational capabilities and

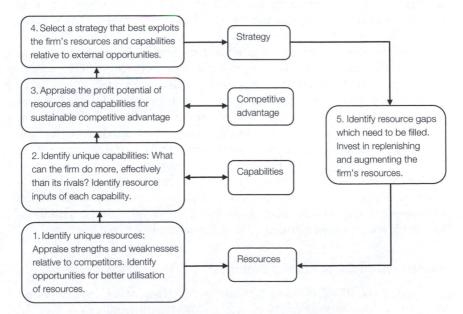

Figure 5.3 A resource-based view (RBV) of strategy (Grant 1995).

resources. The key to competitive advantage then comes from knowing what these are. This turns on its head traditional notions of strategy that assert that an organisation starts with an environment and fits a business to it. Instead it argues that a manager should start with the organisation and find an environment that fits it. As Whittington (1993) points out, this also reverses the primarily externally focused view of strategy to one which is aimed at inward capacity building which invests in people rather than primarily an outwardly market-driven process of pursuing economic opportunities and exploiting people. In other words, the RBV approach to strategy is based on what a firm is distinctively best at doing rather than on adapting to a market analysis. It focuses on the relationships between people, skills, knowledge (both tacit and explicit), and other resources and how they are mobilised, developed and combined into special organisational competencies that provide a sustained source of competitive advantage. This is highly relevant to the labour-intensive construction sector. As Hillebrandt and Cannon (1990) noted nearly two decades ago, strategically, the most important resources to be managed by construction firms are its human resources. To Hillebrandt and Cannon the labour-intensive nature of the construction process involves a constant strategic juggle to match human resources and management skills to a continually changing array of geographically dispersed projects over time. Later, Langford and Male (2001) argued that a construction firm's competitive advantage is especially dependent upon three core capabilities, namely: its relational assets, its reputation and its ability to innovate. Thus it would seem that a RBV of strategy is more appropriate to construction firms because it places human resources and relationships at the centre of a firm's strategy development process.

The identification of relational assets as a critical resource again points to the potential value of the RBV of strategy to construction firms since it reconceptualises Porter's traditional ideas of linear 'value chains' and of stand-alone unified companies operating in relative isolation into business webs, networks and constellations (Cummings 2005). Essentially, as Cummings (p. 244) points out, the RBV represents a postmodern recognition of the 'rhizomatic and relational' nature of organisational life. However, despite its apparent merits, the RBV has struggled to influence strategy development in the construction sector. While the increasing popularity of alliance and partnership-type contracts in some segments of the construction industry might suggest otherwise (Walker and Maqsood 2008), it is doubtful that this is consciously informed by ideas around the RVB of the firm.

Human resources as 'core capabilities'

According to the RBV, the core capabilities that provide any organisation's competitive advantage are those that are valuable, rare, non-substitutable, imperfectly imitable and imperfectly mobile. Quin and Hilmer (1994) argue that to extract maximum competitive advantage, core capabilities should also be: skills and knowledge rather than products and functions; flexible, long-term platforms capable of adaption and evolution; unique sources of leverage within each step of the value chain; limited in number (not normally above five); areas where an organisation can dominate; important to customers and; embedded in organisational systems.

Essentially, the RBV of strategy argues that strategy should be concerned with the identification and nurturing of these unique and inimitable core capabilities and that the labour force is a key source of these. For construction firms this is especially true and the adoption of this view has very different implications for the development of strategy. For example, Prahalad and Hamel (1990) argue that while short-term competitiveness may derive from price advantages and performance advantages of products, in the long term sustainable competitive advantage derives from the ability to build core capabilities that lead to the development of products that customers have not yet 'imagined' through creativity and 'realised' through innovation. This is fundamentally a human activity. Thus from this perspective, management strategies which may be considered valuable from a Porterian perspective because they are focused on economic efficiency, may be considered damaging from a RBV perspective since they can erode core capabilities by breaking businesses up and destroying historic patterns of relationships between individuals and groups which represent a unique source of competitive advantage (the heritage and social capital of an organisation) etc.

In other words, the difference between the traditional and RBV of strategy is that competitiveness is seen to derive from the inner workings of an organisation rather than its products and services and the way that these inner workings facilitate collective learning among its human resources. Prahalad and Hamel (1990) call this collective knowledge, habits, values, skills, relationships, routines and ways of working, one of an organisation's most important 'core capabilities'. In this sense, traditional HR practices provide little source of competitive advantage since they are easy to imitate. Rather, competitive advantage comes from strategic HR practices which focus on aligning human resources with business goals, developing core capabilities and competencies and managing the internal social architecture of an organisation to facilitate interaction between human resources enabling serendipitous learning, spontaneous cooperation and informal interaction outside formal pathways and processes (Mueller 1996). The tacit knowledge that develops out of these types of informal unplanned interactions is a key organisational resource. This is the knowledge that defines an organisation, that cannot be written down and which is a key strategic resource because its intangibility and invisibility makes it impossible for rivals to imitate (Nonaka 1991, Quinn 1992, Lilley 2001). Nonaka (1991) argued that the importance of informal social networks and tactic knowledge as a core capability highlights the critical role of middle managers in strategy formulation. Sitting between frontline managers and senior executives, it is their role to synthesise the tactic knowledge from each and make it explicit in the development of new products and services. Middle managers are the true knowledge engineers in an innovative company.

So the RBV of strategy is one which is built on what an organisation is capable of doing rather than the external needs that the business seeks to satisfy. The RBV sees these capabilities as assets that must be identified, cherished and developed and not taken for granted. It highlights the roles of managers in identifying the critical resources/assets which provide capabilities that create competitive advantage and in deploying them in ways which harnesses that advantage. Managers must combine their resources, both 'soft' (people, knowledge) and 'hard' (buildings, technologies, data etc) in unique ways to produce a 'distinctive capability' and corporate identity that is hard to imitate. Finally, the RBV highlights the role of organisational relationships, structures, cultures, processes, routines and organisational heritage and history in determining how people behave in ways which produce hard-to-imitate and unique outcomes.

Sustaining competitive advantage through human resources

While the RBV offers a valuable alternative 'values-based' view of strategy to the traditional planned approaches offered by the likes of Porter, it is not without its critics. For example, Kodama (2007) argues that in the real world of managers the RBV can be hard to implement because it fails to answer specific questions about how resources are used to enhance competitiveness. Also, the definition of resources is all-inclusive and does not distinguish between those resources which a manager can and cannot control. Finally, there is no firm insight offered into how managers can acquire and develop dynamic capability and how they can best acquire and develop the competencies which provide the basis of competitive advantage. In responding to these concerns, Grant's (1991) work is useful. He recognised these issues and identified four characteristics of any resource that determine the sustainability of competitive advantage. Using this framework, it is possible to use the RBV of strategy to understand the types of human resources which construction firms should be seeking and developing to help achieve this goal. These four characteristics are discussed below.

Durability

Durability reflects the rate at which any resource becomes obsolete. Competitive advantage depends on durable resources which do not become obsolete quickly yet all organisational resources are prone to obsolescence from a wide range of factors. For example, the pace of technological change may reduce the durability of technical resources, interest rates may reduce the durability of capital, and corporate espionage may reduce the durability of intellectual capital. Similarly, the pace of knowledge development and the capacity of people to innovate and adapt to change has the same effect on the durability of human capital. So the secret to sustaining competitive advantage from human resources is to ensure that employees have these types of attributes (inquisitive, innovative and adaptive) by appropriate recruitment and HR development processes which encourage people to learn, develop, think creatively and to innovate. Human resources then need to be supported by organisational cultures, structures and systems which provide people with the openness, flexibility and information they need to do this.

Replicatability

The sustainability of competitive advantage also depends on how easy it is for a competitor to copy another firm's core competencies which in turn depends on the ability to replicate the key resources that create this. In general, competencies that depend on multiple resources are harder for a competitor to copy than single resource-based competencies. The same applies to competencies that are based on complex and unique organisational routines or which require a high degree of resource coordination and tacit knowledge to function effectively. So to sustain competitive advantage from human resources an organisation needs to recruit and retain people with unique capabilities, from not traditional sources (women, overseas, disabled etc) and from liberal disciplinary bases such as the arts, sciences and humanities and to combine them in unique ways to produce a creative and innovative work environment.

Transferability

If key resources can be acquired on special terms then it is difficult for a competitor to erode competitive advantage. On the other hand, if resources can be acquired by competitors on similar conditions then competitive advantage will be temporary. While it is possible to imagine how a firm may obtain special access to resources such as materials at highly competitive rates it is surprising how many firms fail to do this for their human resources. Traditional approaches to recruiting human resources in construction are remarkably narrow, predictable and replicable, so not a source of competitive advantage for most firms. Yet those that are willing to move outside traditional minimum labour agreements and labour pools and explore a more liberal recruitment base drawing from other fields and disciplines and from minority and under represented groups can experience significant benefits over other firms.

Case Study 5.1

Aligning corporate identity and social responsibility at Brookfield Multiplex

Multiplex Constructions, before its takeover by Canadian asset manager Brookfield, was a highly successful privately-owned Australian construction company. It had a reputation for delivering difficult projects on time and on budget, and a defining corporate culture that reflected the entrepreneurial style of its then owner John Roberts. Brookfield Multiplex (BM) Constructions launched its new brand in late 2010 underpinned by the key values of: care, collaboration, integrity and outperformance. These values were identified by BM employees and key stakeholders as best reflecting the unique identity and culture of the current business.

After analysing the performance of the business and 259 individual projects between 1999 and 2007, BM had also identified the value of planning and the direct relationship of maintaining a core of critical common practices, but at the same time being sufficiently agile, less bureaucratic, yet disciplined, in relation to decision-making and strategic thinking.

Differentiation through innovation has always been at the heart of the business and reflects the entrepreneurial spirit of its founder. This approach has enabled BM to broaden its strategic focus to tackle one of the most challenging aspects of the industry and community as a whole – balancing its corporate and social responsibilities to society and its employees.

In dealing with workplace challenges BM has traditionally been responsive to the needs of individual employees. There has always been a dedicated HR function in BM and strategic people solutions are shared by different levels of management throughout the organisation. Such solutions are considered against an important aspect of the company's general strategy that was to enhance the performance of the business by involvement in initiatives that had a positive impact in the wider community.

In researching factors that influenced recruitment and retention, BM considered the issue of how best to align employee's goals with those of the company, while at the same time addressing employee's expectations, in regard to a range of issues that fell under the banner of Corporate Social Responsibility. One issue that had been identified as being of particular significance, due perhaps to the immediate potential for the company to make a difference, was that of social inclusion.

A sharp focus was added to BM's consideration of social inclusion by the onset of the Global Financial Crisis (GFC). At this time, the Australian Building and Construction Industry, along with its counterparts throughout the developed world, seemed to be heading towards a significant increase in unemployment.

As one outcome of their research, BM identified The Salvation Army Employment Plus (TSAEP; Australian Jobs Program) as a leader in employment solutions. After a number of meetings it was determined by both parties, that there appeared to be a strong basis for collaboration. Interestingly, this basis has been confirmed over time, as being the shared values of both organisations and their common focus on helping the socially disadvantaged in the communities in which they do business.

The GFC drove innovation by removing one of the critical problems typically associated with social issues – money. A solution was needed whereby a financial handout was not the solution. Working with the TSAEP, BM considered how it could provide practical support to jobseekers in the Australian, New South Wales Central Coast region. This region had amongst the highest unemployment levels in the State and the type of support considered was at two levels. At the first level (operational) it included providing access to the company's supply chain, work experience, mentoring and a range of other initiatives designed to assist TSAEP in helping its clients become job ready. At the second level (corporate) it was directed to providing the backup and political support required to ensure that outcomes were consistent with and enhanced the BM/TSAEP strategy.

BM and TSAEP identified that they needed a project to be the 'vehicle' for the solution. The Australian Government's response to the GFC was swift, and included a massive stimulus package for the Building Industry called the Building Education Revolution (BER) program, which had a major focus on social inclusion and employment creation. This crisis provided the opportunity for BM and TSAEP to implement their program in order to meet these requirements by providing solutions that were innovative and that could actually make a real difference to unemployment. The BM strategy had proved to be prescient and was totally aligned with the objectives of the stimulus package.

Innovation occurred on a number of fronts, most notably with TSAEP developing a means to streamline the process for BM subcontractors to recruit local people disadvantaged by unemployment (the Total Solutions Industry Solutions Strategy). In its simplest form this involved TSAEP forming alliances with key industry stakeholders and taking responsibility to coordinate the myriad of agencies and service providers involved in assisting the unemployed to find employment. In its more complete form this involved TSAEP working with BM to engage with and educate their subcontractors on the opportunities available to them to fulfil their labour supply needs whilst also helping the long-term unemployed. It is interesting to note that the majority of subcontractors approached were totally unaware of the level of support that was available through accessing the government's Job Services Australia Network, in this case TSAEP, and also interesting to note that many subcontractors were prepared to accept that unemployment solutions required greater community involvement to be successful.

For BM, this partnership with the Salvation Army on the BER project has been a resounding success. The tender requirements for social inclusion were not only met but exceeded. Following the success of the BER, which became the springboard for a number of other substantial projects throughout Australia, BM formally partnered with TSAEP to further explore how both organisations could leverage each other's relative strengths. A number of future projects have been identified through this process some of which have already commenced.

The BM TSAEP Partnership continues to grow. For BM it has become a key feature of the company's corporate social responsibility strategy and as originally envisaged, is offering innovative solutions directed towards improving outcomes for both organisations and benefiting the wider community. It demonstrates how, through a resource-based approach, organisations like BM can achieve competitive advantage by leveraging their unique corporate values, social capital, human resources and capabilities.

Conclusion

In this chapter we have challenged traditional models of strategy and argued for one which recognises the value of human resources as a source of corporate identity and differentiation. We have argued that a RBV approach to strategic management can provide competitive advantage by providing employees with an intimate understanding of, and connection with, an organisation's goals and vision. However, this will require a paradigm shift for many firms in the construction industry since in the RBV approach to strategy, success is not merely measured in financial terms but in broader terms reflecting what an organisation 'does' rather than what it is simply 'worth'. The RBV is also an inclusive and socially responsible approach to strategy which recognises the value of involving non-traditional stakeholders such as supply chain partners, NGOs and communities in the strategy development process. Again, many construction firms fail to do this effectively. Yet this inclusive approach has numerous advantages, most specifically because it facilitates greater employee cohesion and connectivity with strategy which results in a commonly accepted understanding of what an organisation stands for, what it is, what it does and what is important and what is not. It also reduces the potential for conflict within and between groups that may oppose strategy implementation (where many firms fail) and reduces risk exposure by providing more information into decision-making processes.

A key point which emerges from a RBV of strategy is that the resources that an organisation uses in unique ways to create its core competencies do not all need to be owned by that organisation. An organisation only needs access to them when required and to have the internal capability to use them effectively. Thus while resources may be internal or external to an organisation, core capabilities are always internal. This then leads to questions regarding trade-offs between the control and ownership of resources, a choice which depends on the degree to which ownership is necessary to ensure access to and control over critical resources. For example, in the construction industry, the vast majority of construction firms outsource their human resources at operative level to subcontractors, although it has long been recognised that the lack of control over critical human resources has been a source of many problems. However, it has not been widely acknowledged that the loss of identity associated with such a strategy can also detrimentally affect competitiveness. In response, some firms are experimenting with moving back to direct labour and others are exploring the possibility of reducing dependency on labour altogether by offsite manufacturing and prefabrication. These are separate but important issues addressed in later chapters.

References

Aaker, D. A. (1998) Srategic Market Management, Wiley, NY, USA.

Ansoff, H. I. (1965) Corporate Strategy, McGraw-Hill. New York.

Barney, J. (2001) Resource-based theories of competitive advantage: a ten-year retrospective on the resource based view, Journal of Management, 27(4), 643–50.

Chandler, A. D. (1962) Strategy and Structure, Cambridge, MA, MIT Press.

Cheah, C. Y. and Gavin, M. J. (2004) An open framework for corporate strategy in construction, *Engineering, Construction and Architectural Management*, 11(3) 176–88.

Cummings, S. (2005) Re-creating Strategy, London, Sage.

Dainty, A., Green, S. and Bagilhole, B. (eds) (2007) People and Culture in Construction, Taylor & Francis, London, UK.

Daniell, M (2004) Strategy, Palgrave, Macmillan, Basingstoke, Hampshire, UK.

Gooderham, G. (1998) Debunking the myths of strategic planning, The Management Accounting Magazine (Canada), 72 (4), 24–26.

Grant, K. (1996) Towards a knowledge-based theory of the firm, Strategic Management Journal, 17, 109–22.

Grant, R. M. (1991) The resource-based theory of competitive advantage: implications for strategy formulation, California Management Review, Spring, 114–35.

Grant, R. M. (1995) Contemporary Strategy Analysis: Concepts, techniques, applications, 3rd edn, Blackwell, Oxford.

Green, S. (1998) The technocratic totalitarianism of construction process improvement: a critical perspective, *Engineering, Construction and Architectural Management*, 5(4), 376–86.

Green, S. D. (2006) Discourse and fashion in supply chain management, Pryke, S. and Smyth, H. (eds), The Management of Complex Projects, Blackwell Publishing, Oxford, UK, 236–51.

Green, S. D. (2009) The evolution of corporate social responsibility in construction, Murray, M. and Dainty, A., Corporate Social Responsibility in the Construction Industry, Taylor & Francis, London, 24–54.

Hubbard, G. (2004) Strategic Management: Thinking Analysis and Action, Pearson Prentice Hall, Sydney.

Hillebrandt, P. M. and Cannon, J. (1990) The Modern Construction Firm, Macmillan, London.

Kodama, M. (2007) Knowledge Innovation: strategic management as practice, Edward Elgar Press, Cheltenham, UK.

Kumaraswamy, M. M. (1996) Construction dispute minimisation, The Organization and Management of Construction, Langford, D. A. and Retik, A. (eds), London: Spon, 447–57.

Langford, D. and Male, S. (2001) Strategic Management in Construction, 2nd edn, Blackwell Science Oxford, UK.

Lilley, S. (2001) The Language of Strategy, Linstead S. and Westwood, R. (eds) The Language of Organization, London, Sage.

Loosemore, M. (1999) The problem with business fads, Karim, K. *et al.*, Construction Process Re-engineering, *Proceedings of the international conference on construction process re-engineering,* Sydney, Australia, 12–13 July, 355–63.

Loosemore, M., Raftery J., Reilly, C. and Higgon, D. (2005) *Risk Management in Projects*, Spon Press, London, UK.

Loosemore M. and Flanagan R. (2009) Proceedings of the CEO summit, 30 March, University of New South Sales, Sydney, Australia.

Loosemore, M. and Phua, F. (2010) *Responsible Strategy in the Construction and Engineering Industry: Doing the right thing?*, Routledge, London, UK.

Macfarlane, I. (2008) Look beyond greed: a debt-ridden world of leverage made this mess, *Sydney Morning Herald*, Thursday 4 December, 15.

Mueller, F. (1996) Human resources as strategic assets: an evolutionary resource-based theory, Journal of Management Studies, 33(6) 757–85.

Murray, M. and Dainty A. (2009) Corporate Social Responsibility in the Construction Industry, Taylor & Francis, London, 98–119.

National Economic Development Council. 1983, *Faster Building for Industry*. HMSO: London.

Ness, K., (2010) The discourse of 'Respect for People' in UK construction, *Construction Management and Economics*, 28(5), 481–93.

Nonaka, I., (1991) The knowledge creating company, Harvard Business Review, November–December, 96–104.

Pascale, R. T. (1991) Managing on the Edge, Simon and Schuster, NY, USA.

Pearce, D. (2006) Is the construction sector sustainable? Definitions and reflections, Building Research and Information, 34(3), 201–07.

Peters, T. G. and Waterman, R. H. (1982) In search of Excellence: Lessons from America's best run companies, Harper and Rowe, New York, USA.

Pettigrew, A., Thomas, H. and Whittington, R. (2006) Handbook of Strategy and Management, Sage, London.

Porter, M. E. (1980) Competitive Strategy, Free Press, NY.

Porter, M. E. (1985a) Competitive Advantage, Free Press, NY.

Porter, M. E. (1985b) Competitive Strategy: techniques for analyzing industries and competitors, Free Press, NY.

Prahalad, C. K. and Hamel, G. (1999) The core competencies of the corporation, Harvard Business Review, May–June, 79–91.

Quinn, J. B. (1992) Intelligent Enterprise: a knowledge and service based paradigm for industry, The Free Press, New York.

Quinn, J. B. and Hilmer, F. G. (1994) Strategic outsourcing, Sloan Management Review, Summer.

Quinn, J. B. and Voyer, J. (1996) Logical incrementalism: managing strategy formation, Mintzberg, H. and Quinn, J. B. (eds), The Strategy Process: Concepts, contexts, cases, 3rd international edn, Prentice Hall, Sydney, 95–101.

Salaman, G. and Asch, D. (2003) Strategy and Capability, Blackwell Publishing, Oxford, UK.

Sikula, A. (1996) Applied Management Ethics, Irwin, NY, USA.

Starr, R., Newfrock, J. and Delurey, M. *et al.* (2003), Enterprise resilience: managing risk in the networked economy, *Strategy and Business*, issue 31 Spring, 1–10.

Stewart, M. (2009) *The Management Myth*, W. W. Norton and Company Inc, NY.

Teece, D. J., Pisano, G. and Sheun, A. (1997) Dynamic capabilities and strategic management, *Strategic Management Journal*, 18: 537–53.

Thompson, P and Warhurst, C. (1998) Workplaces of the Future, Macmillan, Basingstoke, UK.

Viljoel, J. and Dann, S. (2000) Strategic Management, Longman, Pearson Education, Sydney, Australia.

Walker, D. H. T. and Maqsood, T. (2008) Procurement innovation and organisational learning, Procurement Systems: a cross-industry project management perspective, London, Taylor & Francis, chapter 8.

Whittington, R. (1993) What is Strategy and Does it Matter?, Routledge, London.

6 Occupational health, safety and workers' wellbeing

Helen Lingard

Introduction

Occupational health and safety (OHS) is often regarded as something that must be 'implemented' within organisations. The production of documents, including safety policies, procedures and plans, is often misconstrued as 'doing safety' in organisational life. It is believed that, by following the principles and practices embodied in established OHS management systems, workplace hazards will be controlled to within acceptable levels of risk. Indeed, there are many companies, even in the high risk construction industry context, that perform very well in OHS. However, the attainment of OHS is the result of more than implementing an OHS management system. OHS is the outcome of organising work properly and holding persons with OHS responsibility accountable for their actions. It is very common that organisations in which third party accredited OHS management systems have been implemented continue to demonstrate considerable 'patchiness' in OHS performance, reflecting project-by-project variation in the degree of local effort that is made to deliver OHS outcomes. It is important to recognise that the implementation of an OHS management system will only make a difference if it is accompanied by genuine commitment, will and effort to improve OHS. It is also essential that OHS is integrated into business and project decision-making. OHS should be a primary responsibility of managers at all levels within organisations, from Chief Executive Officers to first-level supervisors, foremen and team leaders. Regarding OHS as a separate management function decouples it from the 'main game' – the operational management of a construction organisation or project. If this connection is not clear, the activities of OHS advisors can appear irrelevant, uncoordinated and even counter-productive, and the performance of an OHS management system will not meet expectations (Gallagher, Underhill and Rimmer, 2003).

Occupational health, safety and workers' well being is a central component of the ethical and responsible management of people. The human impacts of occupational injury and illness include impaired domestic and daily function, strained family relationships, negative psychological and behavioural responses, stress and loss of vocational function. Although they are difficult to quantify, these impacts are considerable and substantially increase the demand for social and health services. In the construction industry, occupational injuries and illnesses are associated with longer than average work absences (Larsson and Field, 2002) and higher rates of permanent impairment than in other sectors (Guberan and Usel 1998). Health is an important resource, contributing to the economic and social wellbeing of workers and their families and responsible and proper management of workers' health, safety and wellbeing is essential to ensuring that construction workers achieve healthy, productive life outcomes.

This chapter deliberately does not focus upon the components of OHS management systems, which are amply described elsewhere. Instead, the chapter addresses some important points of ongoing debate in construction OHS research and practice. First, the chapter examines the argument that there is a natural identity of interest between workers and employers with regard to OHS because workplace injuries are costly to organisations. The role and structure of frameworks regulating OHS are described. Labour market characteristics increasing the vulnerability of construction workers to workplace injury and illness are then explored. Contemporary theories pertaining to antecedents of workplace incidents are briefly described with their implications for attributing professional responsibility for OHS within the construction supply chain. The impact of emergent causes of work-related injury and illness, including long hours, fatigue and work stressors, is also considered in the construction context. The shift in emphasis from efforts to prevent work injury and illness to the positive promotion of workers' health and well being is described and a case study is used to illustrate health promotion interventions implemented at one prominent construction project in Melbourne, Australia. Key organisational issues of managerial accountability, safety culture and leadership are discussed. A second case study highlights the role played by managers' conduct in shaping the local safety climate and injury performance of construction work crews. Key issues of worker involvement in organisational safety processes and trust are also considered in the construction context. The chapter concludes with a discussion of the importance of promoting a healthy and productive workforce for the sector.

Relationship between safety and profitability: myth or reality?

In recent decades, it has become very popular for OHS writers and practitioners to base the OHS case on the cost savings to be made from reducing the incidence of events resulting in death, injury or illness. Unfortunately the 'business case' arguments for OHS have tended to over-shadow discussion of the moral imperative to provide a healthy and safe workplace. This is problematic because in most instances the provision of a safe and healthy work environment requires investment incurring a monetary cost to organisations. An economic rationalist approach would suggest that when this cost exceeds the monetary savings that will be gained from a subsequent reduction in work-related injuries or illnesses, no further investment should be made. As described below, there are substantial problems inherent in this argument. Indeed, it is unlikely that the economically optimal level of investment in the prevention of work-related injury and illness for a single organisation will produce a socially (or morally) acceptable level of OHS risk. Indeed, the notion of what constitutes an 'acceptable' level of OHS risk cannot be understood in terms of a simple cost-benefit analysis undertaken by management. Rather, it must be considered as part of an ongoing and inclusive discussion between workers (and their representative unions), employers and governments.

It is simplistically argued by many OHS writers that the costs of OHS incidents are so great that existing economic incentives can motivate organisations to invest in safe and healthy technologies and work systems. These arguments are often based upon national or aggregated figures. The costs of workplace injury are indeed substantial. For example, in Australia, the National Occupational Health and Safety Commission (2004) estimated that the cost of workplace injuries to the Australian economy for the 2001–02 financial year was \$31 billion. In the UK, the Health and Safety Executive quantified the costs of safety incidents in a number of British industries, including construction. The study found, on an annual basis, the costs of incidents was 8.5 per cent of the price of tendered work in the construction organisation (Cutler and James, 1996).

However, drawing conclusions about enterprise-level incentives to reduce the incidence of workplace injury from these aggregate figures is problematic. As Hopkins (1999) points out, what matters is not how much injuries cost, but who bears the cost of these injuries. The most relevant question is not how much injuries cost the economy but 'how do the costs borne by employers (who are in the best position to control workplace hazards/risks) compare with the costs of prevention?'

An Australian Industry Commission study (1995) reports that employers incur only 30 per cent of the total costs of workplace injury, with the remaining costs borne by workers and the community. Furthermore, the proportion of costs borne by the employer reduces with the severity of an injury. Thus, the Industry Commission concludes that financial incentives for employers to prevent injuries are 'inadequate, particularly for serious incidents' (Industry Commission, 1995, p. 102). More recent analysis reveals that the proportion of injury costs borne by employers may even be as low as three per cent of total costs, with 44 per cent borne by workers and 53 per cent borne by the community (NOHSC, 2004).

Even when costs are considered at the level of a single enterprise, Hopkins (1999) points out that it is insufficient to base the argument that 'safety pays' on the magnitude of the costs of incidents. A rational calculation needs to be based on a reliable quantification of the costs of incidents *and* the costs that would be incurred trying to prevent these incidents. Hopkins (1999) suggests that, in the case of high frequency injuries, cost-benefit assessment may motivate rational employers to invest in prevention, while this may not be the case for the prevention of low probability catastrophic events. Consequently, he cautions against making the assumption that it will always make sense for employers to invest in OHS.

Organisations need to be more cognisant of the problems inherent in reliance on cost-benefit arguments for investment in OHS. In this context, investment decisions should be based upon more robust processes involving the involvement of workers, whose OHS will be impacted by decisions that are made.

OHS regulation

If it is accepted that the natural economic incentive for organisations to invest in OHS may be insufficient to produce 'acceptable' levels of performance, alternative drivers must be considered. One important driver exists in government intervention in the regulation of OHS.

The way organisations treat their workers is an issue that raises ethical concerns (Greenwood, 2002). Rowan (2000) suggests three ethical principles for the management of people, based upon an individuals' right to pursue their own interests. These are: the right to freedom; the right to wellbeing; and the right to equality. Workers have a moral right to be treated fairly and to be provided with a healthy and safe work environment and work conditions that will not harm their general physical or psychological wellbeing. Logically, it is employers who have a concomitant duty to provide a suitably healthy, safe and hazard-free work environment for workers. In recognition of these rights, modern societies have passed legislation to protect workers' health and safety in and arising from work. However, this OHS legislation can take different forms. In many Western countries, a self-regulatory approach has been adopted. The ability of this approach to deliver high standards of OHS has been questioned by some writers (see, e.g., Dawson, Willman Clinton and Bamford, 1988).

Since the 1970s in the United Kingdom (and the early 1980s in Australia) OHS legislation has moved away from a detailed prescriptive model, in which technical solutions to specific hazards were prescribed in detailed specification standards. In these countries, legislation has shifted to a more 'flexible' approach, in which dutyholders are able to decide, for themselves, how to comply with broad-based general duties. This move followed recommendations made by the Robens Committee of Enquiry in the United Kingdom, which argued that dutyholders had difficulty in identifying what their legal obligations were due to the large volume of OHS legislation that had been adopted on a piecemeal basis since the industrial revolution. Although the law was vast and complex, the 'ad hoc' nature of enactment gave rise to gaps in coverage. Additionally detailed laws prescribing technical solutions to OHS hazards were cumbersome, could not keep pace with technological change and prevented dutyholders from finding new and innovative ways to improve OHS. The Robens-inspired legislative reform moved from a specification-standards approach to a more self-regulatory model, based upon employer and employee consultation and workplace decision-making.

The Robens model includes two principal elements: (i) a single umbrella statute containing broad 'general duties' based on the common law duty of care; and (ii) the empowerment of dutyholders, in consultation with employees, to determine how they will comply with those general duties' provisions. The general duties provisions establish a requirement for dutyholders to provide a healthy and safe workplace, but do not indicate to dutyholders *how* to comply, i.e. they are principle-based, rather than prescriptive. The practical method of compliance with this type of legislation is to be determined by dutyholders. Moreover, the general duties are not absolute and are limited by words like 'so far as is practicable' or 'reasonably practicable.'

In some jurisdictions, this legislative approach establishes a three-tiered 'hierarchy' in which the general duties contained in the 'umbrella' Act are supplemented by process-based regulations and hazard-based (non-mandatory) codes. Process-based standards go further than establishing general duties and focus attention on how OHS is to be managed. They require dutyholders to follow a certain process, or series of steps in the identification, assessment and control of workplace hazards. Non-mandatory codes of practice provide a greater level of detail to dutyholders about ways in which they can comply with their general duties in relation to specific hazards, e.g. working at height and manual handling. Though not mandatory, codes assist dutyholders by suggesting *how* they are able to comply with the legislation on a particular issue. Codes of practice possess quasi-legal status, as measures described in a code are deemed to comply with the legislation (Bluff and Gunningham, 2003).

In contrast to the UK/Australian legislative model, some countries have retained a large volume of prescriptive OHS law. For example, in the USA, the Occupational Safety and Health Act (OSHA), is the primary federal law governing OHS. Section 5 of OSHA contains a general duty clause requiring employers to: (i) maintain conditions or adopt practices reasonably necessary and appropriate to protect workers on the job; (ii) be familiar with and comply with standards applicable to their establishments; and (iii) ensure that employees have and use personal protective equipment when required for safety and health. However, OSHA also contains a great deal of detailed prescriptive specification standards establishing requirements for the control of mechanical and chemical hazards. International comparative figures produced by the International Labour Organization suggest that the OHS performance of the US construction industry is notably poorer than that of the UK and/or Australia. For example, the ILO LABORSTA database reports a fatality rate of 4.4 per 100,000 workers for the construction industry in Australia in 2008, compared with a rate of 10.0 for the USA in the same year. The latest figures available in the LABORSTA database for the UK relate to 2006, when the fatality rate for construction was 4.5 per 100,000 workers. However, it is not possible to attribute these differences to the legislative framework as there are many other differences (e.g. in the prevailing economic framework and industrial climate), that are likely to be relevant.

Vulnerable workers

One assumption underpinning the Robens Committee's thinking was that there is a 'natural identity' of interest between employers and workers in relation to OHS. Nichols (1997) argues that this premise cannot be uncritically accepted. If a 'natural identity' exists then Nichols (1997) questions why levels of workplace injury are so persistent as the only explanation can be that workers and/or employers consistently act irrationally in failing to pursue their own interests. As the above discussion of the costs of injury, illness and death reveals, it is dangerous to assume that employers acting rationally will invest in prevention. Indeed, Nichols (1997) argues that OHS must be understood in relation to the larger economic system in which workers and employers operate. In this system, workers' unsafe acts are sometimes a 'rational' response to production pressures. The self-regulatory focus of Robens thinking, in which employers and workers would work together to find solutions to OHS problems through a process of consultation at the workplace may actually be misguided. Workers' ability to influence OHS standards in the workplace is likely to be impeded by the power imbalance inherent in the employment relationship. Further, the ability to engage in *meaningful* consultation with employers in relation to OHS is likely to be particularly difficult among vulnerable worker groups, which make up a reasonable proportion of the construction industry. What follows is a brief discussion of three groups of vulnerable workers whose labour construction organisations are increasingly reliant upon, namely young workers, migrant workers and workers engaged in casual or precarious employment.

Young workers

The Australian construction industry has a relatively young age profile. For example, Australian Bureau of Statistics data indicates that 62.6 per cent of construction workers are under the age of 45 compared to 59.6 per cent of workers in all industries. The percentage of construction workers 45 years or over was 37.4 per cent in construction compared to 39.8 per cent in all other industries (Government of South Australia, 2009). Notwithstanding this age profile, the physical nature of some trade sectors of construction work can result in an earlier age transition from trades to other roles within the construction industry, which impacts upon the supply of labour in these trades. It is therefore imperative that the work health, safety and wellbeing of younger construction workers is carefully managed.

The work injury rates of young workers, aged 15–19 are reported to be significantly higher than those of older adults in European and US research (Salminen, 2004; Breslin and Smith, 2005). For example, in a study of fatal incidents in the US construction industry involving contact with electricity, Janicak (2008) reports a significantly higher proportion of fatalities in the 16 to 19 year age group. This group had an actual Proportionate Mortality Ratio (PMR) of 20, compared to an expected PMR of 13.8. Lavack, Magnuson, Deshpande, Basil, Basil and Mintz (2008) report that 50 per cent of accidents involving young workers occur during the first six months in a job.

In Australia, the rate of work-related death among workers aged between 15 and 19 was 3.4 per 1,000 between 1989 and 1992. The majority of the 106 work-related deaths to young workers occurring in this period involved labourers, tradespersons (many of whom are apprentices) and machinery operators, i.e., occupations commonly found in the construction industry (Driscoll, 2006). Further, construction was one of the main industries in which young Australian workers were killed (with agriculture, manufacturing and trade). Indeed, construction accounts for 19 per cent of the work-related deaths of workers under 19 years of age, but only 13 per cent of all work-related deaths, which is indicative of the vulnerability of young construction workers.

Reasons for the prevalence of work injury among young workers are complex and include personal factors, such as inexperience, levels of emotional and physical maturity and cognitive functioning. However, young workers are often employed in low paid or part-time jobs on a temporary, casual or informal basis. They may not know their legal rights, or be lacking in the self-confidence, communication or social skills to express safety concerns (Loughlin and Frone, 2004). Chin, DeLuca, Poth, Chadwick, Hutchinson and Munby (2010) suggest that, in Canada, although OHS training is a legal requirement, up to 50 per cent of young workers do not receive any OHS training. Further, the training that is provided is typically informational, providing young workers with knowledge of OHS hazards. Rarely is OHS training for young workers 'instructional' in the sense that it provides young workers with a deeper understanding of the reasons for unsafe conditions in the workplace and/or skills they need to address the social risk associated with making a safety complaint in their workplace (Chin *et al.*, 2010).

Even if they understand their legal rights, young workers may be eager to earn the approval of their supervisors. Tucker, Chmiel, Turner, Hershcovis and Stride (2008) report that young workers are less willing or able to engage in communication aimed at improving workplace safety (a term they refer to as employees' 'safety voice') than older workers. The researchers surmise that this may be due to a desire to appear hardworking and able to carry out allocated work tasks. The reluctance to voice safety complaints or concerns is particularly prevalent among male workers in industries including construction (Breslin, Polzer, MacEachen, Morrongiello and Shannon, 2007). In particular, young male workers are reported to accept workplace injury as 'part of the job' about which there is no point complaining. Breslin *et al.* (2007) suggest that this attitude is fundamentally linked to the endemic industry culture and prevailing power imbalance that exists in many workplaces, particularly when young workers are employed on a casual, informal or temporary basis. In this context, young workers have little control over their work environment and their subordinate position leads them to accept unsafe conditions and even injuries.

Migrant workers

Globalisation has seen an increasing use of migrant workers in the construction industries of many countries. In the USA, the number of Hispanic workers in the construction industry has tripled over a ten-year period to reach three million in 2006, almost a quarter of the industry (Dong, Fujimoto, Ringen and Men, 2008). In one UK study by Bust, Gibb and Pink (2008), the average proportion of migrant workers employed by contracting organisations was ten per cent but one contractor indicated migrant workers made up 40 per cent of the workforce. In Australia, Non-English-Speaking Background (NESB) migrants account for approximately 12.5 per cent of the nation's 800,000 construction workers (Trajkovski and Loosemore, 2006). The increasing reliance on foreign construction workers presents significant challenges for the management of OHS. Research suggests that migrant workers in many countries are at a higher risk of workplace death, injury and illness than domestic construction workers. For example, Dong *et al.* (2008) report that between 1992 and 2006, Hispanic workers in the USA consistently experience a higher rate of fatal falls than non-Hispanic workers. Further, during this period, the death rate of Hispanic workers increased while it remained the same for non-Hispanic workers. Although Hispanic workers were 1.5 times more likely to die in a fall than non-Hispanic workers, the rate of falls was particularly high among Hispanic workers born outside the USA (5.5 compared to 3.8 per 100,000 FTE for all construction workers and 4.1 per 100,000 FTE for Hispanic workers born in the USA). In Australia, Trajkovski and Loosemore (2006) cite Australian Bureau of Statistics data showing injuries to foreign-born workers make up 29 per cent of documented workplace grievances. Many of these injuries involved tradesmen or labourers.

One of the most immediate challenges associated with managing a culturally diverse workforce is the need to communicate OHS information in a manner in which it can be understood and acted upon (Bust *et al.* 2008). Trajkovski and Loosemore (2006) suggest that, in Australia, the English language proficiency of many foreign construction workers has not been adequately addressed in the design and delivery of mandatory OHS training programmes. Further, site-based communication of safety requirements and instructions is rarely communicated in a language other than English. Bust *et al.* (2008), advocate the use of symbols, photographs and videos to communicate OHS messages, but point out that it is essential to evaluate how OHS messages are received and understood by workers of different ethnicities and/or cultures. Another important aspect of addressing the OHS challenge presented by migrant workers is in finding ways to secure the participation of those workers in OHS management processes. Haslam *et al.* (2005) identify worker participation in managing OHS as important in the identification of solutions to workplace OHS problems. There is evidence that migrant workers are less likely to engage in employer-employee consultation concerning OHS due to their often limited formal education, language difficulties, position of economic need and low level of knowledge about OHS legislation (Williams, Ochsner, Marshall, Kimmel and Martino, 2010). Williams et *al.* (2010) describe a peer-led participatory OHS training programme that was developed and trialled among Hispanic workers in the US construction industry. The training programme was implemented in partnership with unions and community groups and was designed to increase OHS knowledge and capability by providing group training to 'day labourers', a particularly vulnerable group of workers in the USA. A series of focus groups were initially conducted with representatives of the target population to identify typical exposures and work situations. Training activities were developed based upon principles of 'active learning' and training in topics, such as falls prevention and legal rights, was delivered through 'peer trainers.' Williams *et al.* (2010) report increased OHS knowledge among participants following the training, as well as an increased willingness among participants to communicate OHS information to their co-workers and advocate workplace safety.

Casual (and often precarious) employment

In developed countries, there has been a shift away from permanent full-time employment to an increasing reliance upon temporary, part-time, subcontracting, labour hire and own-account self-employment (Fabiano, Currò, Reverberi and Pastorino, 2008). Research suggests that job insecurity associated with precarious forms of employment can have a deleterious effect on workers' physical and psychological health and wellbeing (Quinlan, 2007; Emberland and Rundmo, 2010). In an analysis of Italian OHS data, Fabiano *et al.* (2008) point out that the frequency index of workplace accidents is 2.55 times higher for temporary workers in all industries than it is for the highest risk sector at a national level (which was the building industry). They conclude that temporary work may be considered a risk factor for workplace accidents and suggest that this may be explained by levels of industry or organisation-specific experience and/or knowledge, insufficient training, a high workload and work pressure. Similarly, in Spain, Guadalupe (2003) demonstrates that workers engaged in fixed-term contracts have a significantly higher probability of having an accident in the workplace than workers on permanent contracts.

Subcontracting is a key feature of the construction industry, which is known to present significant challenges in the management of OHS (Arditi and Chothibongs, 2005; Loosemore and Andonakis 2007). Australian research by Mayhew, Quinlan and Ferris (1997) reveals that the 'payment-by-results' system under which most subcontract work is undertaken pushes contractors to work excessive hours and 'cut corners' with respect to OHS. Further, organisational complexity and ambiguity about OHS responsibility, associated with 'pyramid' subcontracting, inadequate regulatory controls and the low levels of union representation among the self-employed (also discussed in Chapter 10) increase the level of OHS risk experienced by subcontracted/self-employed workers. Another worrying aspect of the prevalence of 'own account' self-employed workers in the Australian construction industry is that injuries to these workers are not reflected in national compensation-based OHS statistics. Mayhew *et al.* (1997) point out that compensation data are used to inform preventive strategies so the absence of injuries to these workers from compensation statistics means that self-employed subcontractors are seldom the focus of prevention efforts. Because they are ineligible for workers' compensation, financial pressures can result in self-employed workers not seeking proper treatment and/or continuing to work while injured, increasing their risk of developing long-term, chronic conditions.

In the construction industry it is evident that age, work inexperience, contingent work arrangements and ethnicity interact in complex ways to render specific groups at considerable risk of work-related harm. In addressing the OHS challenges it faces, the construction industry must carefully consider how vulnerable worker groups are socialised into industry practice to ensure that poor OHS standards and work injury are not regarded as the 'norm.' Efforts to educate vulnerable worker groups about OHS risks and their rights under OHS legislation are critical in order to provide the knowledge-base for these workers to advocate for their own and others' OHS. However, as Breslin *et al.* (2007) note, placing responsibility primarily on vulnerable worker groups to look after their own OHS is insufficient because imbalanced power relationships in many workplaces are likely to discourage workers from making safety complaints. This problem will be compounded for vulnerable workers. It is therefore important that organisational human resource management policies and procedures are designed to redress these inherent power imbalances and actively engage all workers in making decisions about OHS in the workplace. The provision of specific OHS training to vulnerable worker groups, the implementation of culturally appropriate orientation programmes for migrant workers and/or special initiatives to mentor young workers are possible strategies to be considered, within the broader HRM functions of workforce recruitment, development and management.

Nature and antecedents of injury

Injury/incident causation

Injuries and incidents can be explained in different ways depending upon the causation model that is used to analyse them. Different models of injury/incident causation emphasise different aspects and are likely to give rise to different recommendations for prevention (Katsakiori, Sakellaropoulos and Manatakis, 2009). Lundberg, Rollenhagen and Hollnagel (2009) refer to this phenomenon as 'What-You-Look-For-Is-What-You-Find' and suggest the corollary of this is 'What-You-Find-Is-What-You-Fix.' It is therefore very important to comprehend the assumptions embedded in different theories about how factors interact to cause injuries/incidents.

Early models of injury/incident causation were simple linear 'cause-effect' models that described an incident as a sequence of events that occurred in a specific order, such as the 'domino model' (Heinrich, 1959). An injury was seen as the logical conclusion of a sequence of events that commenced with a person's ancestry and social environment. This was a precursor to personal factors which, in turn, contributed to the existence of unsafe acts and/or conditions. Unsafe acts and conditions were the immediate precursors of accidents, which gave rise to injury. Heinrich's model has been criticised for focusing too much attention on the injured worker and the immediate circumstances surrounding accidents.

More recent models of injury/incident causation have recognised that injury/incident causation is more complex than the early linear sequential models would suggest. In fact systemic causative factors that may be present in organisations for many months or even years without the occurrence of an incident are now understood to be important antecedents to injury (Lundberg *et al.*, 2009). So-called 'systemic' causation models highlight organisational and cultural factors in creating the conditions in which a precipitating event can result in a major incident. James Reason's 'Swiss cheese' model is the most widely cited model of this type. According to Reason (1990), incidents are caused by a complex interaction of latent and active failures. Active failures are immediate observable causes, similar to Heinrich's unsafe acts or conditions. These can be easily identified. However, latent failures may also be present in work systems. In a sense these are 'accidents waiting to happen.' These can include poor design, low levels of training, a mismatch between levels of competence and responsibility and other systemic deficiencies. Over time, work systems build up defences against these latent failures. In local workplaces, latent conditions combine with natural human tendencies and result in human errors or violations.

These are unsafe acts committed at the human-system interface. Reason suggests that many unsafe acts occur, but very few of them result in losses because systems have in-built defences, likened to layers of Swiss cheese. But, like Swiss cheese, these barriers have holes in them which vary in size over time. Should a situation arise in which the holes 'line up', the system's defences fail and errors result in organisational accidents. Systemic models of injury/incident causation permit an analysis of causal factors that are chronologically, geographically or organisationally removed from the worksite. Thus, the focus is not solely on the immediate circumstances surrounding the incident. In the construction context, this means that the cause of injuries/incidents may be traced back to systemic failures in the way that construction projects are procured, organised and managed (Suraji, Duff and Peckitt, 2001; Manu, Ankrah, Proverbs and Suresh, 2010). An implication of adopting this systemic approach is that the investigation of workplace injuries/incidents in construction should examine the contribution of all parties, beginning with the injured worker, through supervisors, site managers, construction planners, construction company senior managers, designers, project managers and even clients.

A report prepared by Loughborough University and UMIST on behalf of the UK's Health and Safety Executive sought to test a systemic model of OHS incident causation by carefully investigating the causes of 100 construction incidents. The research team used the information obtained from people involved in selected incidents, including the victims and their supervisors, to describe the processes of incident causation in construction (HSE, 2003). The HSE model identifies originating influences affecting incidents in construction as including client requirements: features of the economic climate; the prevailing level of construction education; design of permanent works; project management issues; construction processes; and the prevailing safety culture and risk management approach. In particular, the analysis of the 100 incidents revealed that, in half of the cases, the risk of accidental injury could have been reduced with a change to the permanent design of the facility being constructed. Deficiencies in the risk management system were also apparent in almost all of the 100 incidents studied, which represents a significant management failure. Project management failures were commonly reported, most of which involved inadequate attention to coordinating the work of different trades and to managing subcontractors to ensure that workers on site had the requisite skills to perform the work safely. The next level of contributing causes identified in the HSE model is termed 'Shaping factors' which include issues such as the level of supervision provided, site constraints, housekeeping and the state of workers' health and fatigue (discussed in more detail below). Poor communication within work teams was also identified as an important shaping factor. The most immediate circumstances in the HSE incident causation model are the suitability, usability and condition of tools and materials, the behaviour, motivation and capabilities of individual workers and features of the physical site environment, such as layout, lighting and weather conditions. While it is important to identify these immediate circumstances, the model acknowledges that construction incidents occur as a result of a complex process, involving proximal, as well as distal causes 'upstream' of the construction site. This presents a challenge for human resource management, as it requires OHS management activities be implemented across an inter-organisational landscape. In the management of OHS, the human resource function cannot be solely internally focused and requires systems and behaviours that support liaison and coordination with a wide range of external organisations with which a construction firm does business.

The HSE model is particularly helpful in the analysis of injuries/incidents in the construction context because it adopts a similar framework to that presented by Reason, but it also reflects the construction industry context. Thus it supports an analysis of human, technological and organisational factors in project procurement and site-specific environments and has considerable potential to direct prevention efforts.

Long hours, fatigue and burnout

One causal factor in construction safety incidents that has received insufficient attention until recently but is a key feature of contemporary models of injury causation is fatigue. It is widely reported that fatigue results in impaired performance and 'accidents.' (Williamson, Lombardi, Folkard, Stutts, Courtney, Connor, 2011). Given the long work hours and physically demanding nature of construction work, the impact of fatigue on the OHS of construction workers warrants attention.

Construction workers tend to work long hours. For example, in Australia, Lingard and Francis (2004) report that the average number of hours worked each week is 63 among site-based employees in direct construction activity, 56 hours among employees who worked mostly in site offices and 49 among employees in the head office of construction companies. The number of hours worked each week is a significant predictor of worker burnout (Lingard and Francis, 2005). Burnout, itself a form of diminished psychological wellbeing, has also been associated with the experience of distress, anxiety, depression, reduced self-esteem and substance abuse (Maslach, Schaufeli and Leiter, 2001).

Construction workers may be at a greater risk of injury as a result of the hours that they work because long hours are linked to elevated risk of error and injury (Olds and Clarke, 2010). Dembe *et al.* (2005) report that, for every five hours worked past 40 hours per week, the average risk of injury increases by 0.7 injuries per 100 worker hours. Further, working more than 60 hours increases the risk of injury by 23 per cent (Dembe *et al.* 2005). Working long hours increases exposure to work hazards, such as noise, chemicals, physical demands, heat and psychological stress (Caruso *et al.* 2006). In construction these exposures (and their effects) are likely to vary by trade. For example, in a comparative analysis of the fatigue and health effects experienced by construction trades, Chang *et al.* (2009) report that scaffolders have shorter sleeping hours and higher levels of physical pain than other trades. In the same study, concretors, who tended to be older than other worker groups, experienced diminished right-hand grip strength and back strength after a workshift, which the researchers attribute to posture and the vibration effects associated with concreting tasks (Chang *et al.* 2009). A recent analysis of fatigue experienced by workers at forty construction projects in Queensland, Australia revealed that average weekly work hours and commuting time were significant correlates of fatigue. Fatigue levels were also related to a range of wellbeing indicators including sleep disturbances, headaches, gastro-intestinal problems, respiratory problems, overall physical illness and psychological wellbeing (Hobman *et al.* 2010). The rate of 'near misses' was significantly greater among workers reporting high fatigue levels.

Long hours also reduce time available for sleep and recovery, which is essential for workers' wellbeing (Sonnentag and Zijlstra, 2006). Research has linked workplace sleepiness with self-reported safety behaviour, such that people who experience high levels of workplace sleepiness report low levels of safety behaviour (DeArmond and Chen, 2009). Workplace sleepiness was also positively linked to the frequency and severity of pain experienced by workers. The link between sleep and safety is evidenced in research by Barnes and Wagner (2009), who report that changes to daylight saving reduces average sleep by 40 minutes per night and is associated with a significant increase in both the frequency and severity of workplace injuries.

Folkard and Lombardi (2006) argue that regulation that restricts the total number of hours worked per week, such as that implemented in Europe under the EU's 'Working Time Directive', may not be the best solution to the problem, because injury risk is not just a function of the number of hours worked per week but the configuration and length of shifts that are worked, as well as the frequency and duration of rest breaks. The length of shifts appears to be a risk factor. Folkard and Lombardi (2006) report that relative to eight-hour shifts, ten-hour shifts are associated with a 13 per cent increased risk of injury and 12-hour shifts with a 27.5 per cent increased risk of injury. The risk of injury also increases progressively over successive workshifts, although this increase is more dramatic in successive night shifts than successive day shifts. Folkard and Lombardi (2006) report that the risk of injury increases substantially and linearly as the length of time between breaks increases. Thus, frequent breaks can apparently negate some of the effects of shift length.

From a human resource management perspective, there needs to be a better understanding of the implication of work shift design and rostering arrangements for OHS in the construction context. In particular, the question of sufficient rest and recovery within work schedules warrant specific investigation. However, any redesign of work schedules in the industry will inevitably involve interaction between HR and other construction management functions, including cost planning, programming and resourcing of construction work.

Workers' health and wellbeing

Hillier *et al.* (2005) write of an 'endemic un-wellness' that is affecting employees' behaviour within organisations, suggesting that a large number of employees and, by logical inference, organisational cultures are unwell. This is compounded by an ageing workforce. For example, 33 per cent of Australia's 864,100 construction workers is aged between 45 and 70. Of those, 83 per cent suffer from health conditions classed as national Health Priority Areas, i.e., arthritis or osteoporosis and musculoskeletal conditions, asthma, cancer, cardiovascular disease, diabetes, injury, mental health and obesity (ABS, 2008). This is a concern because the health of a workforce is essential to productivity, performance and efficiency (Miller and Haslam, 2009). While it has long been recognised that workplaces expose workers to physical and chemical hazards, researchers have only recently begun to expose the health impact of long hours and psycho-social stressors. Another recent development has been the blurring of the distinction between occupational and non-occupational health effects (Drennan, Ramsay and Richey, 2006). There is a growing recognition that employee health and wellbeing are influenced by a complex interaction of factors in work and non-work domains.

In several countries, projections of future health care have led policymakers and some large companies to integrate preventive OHS management activities with positive health promotion activities. This integrated approach arguably optimises health outcomes through the simultaneous assessment and reduction of both occupational and lifestyle health risks.

A report published by Pfizer Inc in 2001, titled '*The Health Status of the United States Workforce*' provided a summary of US workforce health data. According to the report, on average, employees with hypertension miss 67 per cent more workdays per year than non-hypertensive workers (Peregrin, 2005). Although hypertension is asymptomatic, Peregrin (2005) identifies serious behavioural risk factors associated with this condition, including smoking, sedentary lifestyle and obesity. In the light of this report, in the United States, NIOSH introduced a campaign called 'Steps to a Healthier US Workforce.' Hillier *et al.* (2005) describe a similar initiative called the 'Healthy Workforce Initiative' in the UK. This initiative was jointly sponsored by the Department of Health and the principal OHS regulatory body (the Health and Safety Executive) thereby bringing workforce health issues into the mainstream OHS arena.

In Australia a national telephone survey of 16,304 Australian workers conducted between 1998 and 2001 revealed that a large proportion of Australian workers rate their health to be sub-optimal (Korda *et al.* 2002).

Male, blue-collar workers rated their health particularly poorly compared to other groups. These differences persisted after controlling for confounding variables including age, smoking and employment intensity (Korda *et al.* 2002). Almost two-thirds of the Australian workers sampled reported a current long-term health condition, such as asthma, arthritis, hayfever, back pain, cardiovascular disease or other long-term health conditions. Regular physical activity has been found to lower the risk of heath-related disease such as cancer (American Institute for Cancer Research, 2007) and positively impact on work-life balance, health and wellbeing. Studies have indicated that construction-based blue-collar workers are less likely to engage in leisure-time physical activity (Burton and Turrell, 2000), which may be due to levels of physical labour associated with work (Sorensen et al. 2009).

Health promotion

An increasing number of organisations have initiated programmes designed to improve the general physical and mental health of their employees (DeGroot and Kiker 2003). DeGroot and Kiker (2003) distinguish between reactive employee health management programmes and those focused on more positive health promotion. Reactive programmes are those in which assistance is only provided once a particular health problem, for example alcoholism, is identified and help is sought. In contrast, occupational health promotion programmes focus on changing behaviours, at work and outside work, *before* adverse health outcomes occur. These programmes are becoming more widespread in construction (see below) and are designed to promote behaviours that will improve employees' fitness, health and general wellness. Typical components are:

- the provision of nutritious options in cafeterias and vending machines and/or employer subsidies for the purchase of healthy foods (which are often more expensive than non-healthy foods);
- smoking cessation interventions;
- encouraging exercise by subsidising memberships of fitness centres and by demonstrating management commitment to exercise; and/or
- offering onsite health education and screening for issues like high blood pressure or cholesterol (White 2005).

One interesting and unresolved aspect of health promotion initiatives relates to the extent to which organisations should 'interfere' with workers' lifestyle choices outside of work. This point is particularly contentious in relation to 'fitness for work' programmes that involve drug testing of workers (widely used in the Australian mining sector). In order to justify the implementation of such programmes, employers are required to demonstrate a clear connection between workers' activities out of work hours and their ability to perform their work during work hours. Further, the privacy of workers must be very scrupulously safeguarded in the implementation of these programmes. However, in relation to programmes designed to promote more general health and fitness, it is equally important that these programmes be operated in a fair, reasonable and non-discriminatory manner. As in all other aspects of the management of workers' health, safety and wellbeing, the involvement of workers in the design of health programmes is strongly recommended.

Case study 6.1

West Gate Freeway upgrade, Melbourne, Australia

The West Gate Freeway upgrade project aimed to eliminate conflicting merging and weaving movements along Melbourne's most heavily trafficked and economically important transport connection by constructing extra collector–distributor lanes in both directions. The freeway was widened by one lane, a new elevated carriageway was constructed and a major interchange was redesigned. The length of the 5.5 kilometre stretch of freeway covered by the project was a mixture of at grade and elevated carriageways and, during construction, work was required to take place adjacent to the existing freeway that remained open, as well as above other roads, railways and tramlines. The requirement to minimise traffic disruption necessitated construction work at night and during weekends. The project was delivered using an innovative alliancing delivery mechanism. Alliance participants included VicRoads, Thiess, Baulderstone, Parsons Brinkerhoff and Hyder Engineering.

'Health and Wellbeing' was established as one of the project's Key Result Areas, for which all managers were given responsibility and accountability. Early in the life of the project a Health and Wellbeing Committee was established with responsibility for the development of health and wellbeing strategies and monitoring the project's performance in this important area. During the life of the project, a range of initiatives was implemented. These included the provision of onsite flu vaccinations, cholesterol and blood pressure testing and workstation ergonomic assessments for staff based in the project office. To help workers relax during work breaks, books, puzzles and magazines were provided in lunch rooms and a table tennis table was installed in a designated recreation area onsite. Following a project worker's prostate cancer diagnosis, a special seminar was arranged and delivered by a leading Melbourne medical specialist on this topic. A similar information session was conducted in relation to skin cancer prevention and onsite skin cancer checks were carried out. Activities designed to promote a healthy lifestyle among workers included a 'quit smoking' seminar and the provision of onsite advice from a dietician. Various initiatives designed to promote physical activity were implemented, including the sponsorship of workers who joined a 'Step Challenge' exercise programme and who entered charity sporting events. The project also funded bicycle repairs and encouraged workers to participate in a 'ride to work' day. To combat issues of work stress and long hours, stress and time management workshops were held.

During November and December 2009, two work-life evaluation workshops were held with workers of the West Gate Freeway upgrade project. The first workshop was held with seven salaried (professional/managerial) workers. The second workshop was held with seven waged (blue collar) workers. The workshops explored the effectiveness of health and wellbeing strategies implemented at the project. The workshops indicated that the health and wellbeing strategies had been positively received by both salaried and waged workers at the project. One participant commented: '*I think they [the strategies] cater for everybody because there's a lot of people at our depot that were at the prostate cancer seminar and skin cancer checks and even the health checks I heard yesterday was right across the whole board, so it was pretty good*'. Participants also commented favourably on the peace of mind that participating in health-checks provided them with. For example, one said '*I mean with all the health stuff for me it was stuff that I wouldn't have done otherwise. It was just good to do it just for peace of mind*'. Another remarked '*It's a peace of mind thing, quite a few of those things [strategies], okay you know you're fine, you've got a few minor problems, but you can get them sorted out if you know what they are. And I think that's good because it takes the worry away, the stress of it and allows you to focus on your work.*'

Work stress

The construction industry is a high risk industry for work stress. In an analysis of stress among construction site managers, Sutherland and Davidson (1989) identify inadequacy of information flow, onerous paperwork and excessive workload as the top three stressors. Leung *et al.* (2007) also report high levels of objective stress in construction estimators, associated mainly with a perceived lack of autonomy and/or low levels of reward. In a study of Hong Kong construction industry employees, onerous bureaucracy, a lack of opportunity to learn new skills and work–family conflict were ranked the three most difficult stressors to manage (Ng, Skitmore and Leung, 2005).

Work stress is associated with lower levels of job performance. For example, Djebarni (1996) reports that the relationship between leadership and effectiveness is contingent upon levels of stress and provides some evidence for a curvilinear relationship between stress and performance, i.e., performance is low in conditions of low and high stress, but high when stress levels are moderate (the inverted U-curve). However, Leung *et al.* (2008), report that the task performance of construction project managers is inversely linked to objective stress and found no evidence for a curvilinear relationship. They attribute this to the fact that the objective stress levels of project managers are consistently above the threshold at which the inverted U-curve would apply. More recently, Leung *et al.* (2010) report emotional stress is positively linked to workers' injury experience in the Hong Kong construction industry. In Australia, Haynes and Love (2004) identified workload, long hours and insufficient time with family as the three most significant stressors experienced by construction project managers. They suggest that the sources of stress have significant implications for the adjustment or coping mechanisms available to project managers, who will have limited access to social support outside the workplace to help them deal with stress. Also in Australia, Love *et al.* (2010) report that employees of contractors suffer higher levels of stress and lower levels of workplace support than consultants, suggesting that workplace support can help to mitigate the harmful effects of high-stress work. Increasing research evidence supports an ecological model of stress, in which work and non-work environments interact in complex ways to shape the experiences of construction industry workers (see, for e.g. Leung *et al.* 2008).

Stressful work conditions are serious risk factors for employees' mental health but it is now known that stress is also a risk factor in musculo-skeletal disorders (MSDs). This relationship is strong both in relation to the *frequency* and *duration* of MSD claims. For example, *stressors* or *sources of stress* play an important role in the development, exacerbation and maintenance of work-related upper extremity disorders (Bongers *et al.*, 2002) and back pain (Linton, 2001). Indeed, Linton (2001) suggests that eliminating stress risk factors could reduce the number of cases of back pain by up to 40 per cent. The pathways by which work stress contribute to MSDs are not fully understood but Bongers *et al.* (2002) suggest that stressful work conditions: (i) have a direct effect upon the speed of movements, applied force and/or posture; (ii) cause physiological changes leading to MSD problems; (iii) lead to a different appraisal of MSD problems; and (iv) influence the transition from acute to sub-acute to chronic MSD pain.

One form of work stress experienced acutely by construction workers is work–family conflict. Lingard, Francis and Turner (2010) report that Australian construction workers experience higher levels of work–family conflict than those reported by workers in a variety of other industries. A longitudinal study by Jansen, Kant, Amelsvoort, Kristensen, Swaen and Nijhuis (2006) shows a clear correlation between work-family conflict and sickness absence, even after controlling for other issues, such as workers' age and the presence of a long-term disease. Research also reveals that work–to–family conflict acts as the mechanism by which adverse work conditions translate into depression (Franche *et al.* 2006). According to Wang *et al.* (2007) work-family conflict is significantly associated with mental disorders in the American working population. This association was found for both women and men, although the association was stronger in men aged between 26 and 45 years of age and among married or divorced men with children. The latter finding is pertinent to the predominantly male construction industry. Wang *et al.* (2007) suggest this might be due to the fact that middle age is a period of high productivity in which many workers also start a family. The combination of pressures to provide financially and participate in family life at this busy time, the researchers suggest, takes its toll on men's mental health. However, there is also evidence that women in construction occupations are also badly affected by work-family conflict Sang *et al.* (2007) compared the experiences of male and female architects in the United Kingdom and report that female architects experience significantly lower overall job satisfaction and significantly higher levels of work–family conflict and turnover intention than their male counterparts.

Work-family conflict also impacts upon health and wellbeing indirectly, via employees' health-related behaviours. For example, Allen and Armstrong (2006) report that family interference with work is associated with the consumption of more fatty foods and less physical activity, while work interference with family is associated with lower consumption of healthy foods. Research has linked work–to–family conflict and role overload with unhealthy food choice coping strategies, for example eating take-away or fast food rather than home-cooked food, suggesting that this has serious implications for the nutrition and health of working parents and their children (Devine *et al.* 2006).

There is now a strong body of evidence to suggest that experiences at work can spill over into workers' activities outside of work and, where this 'spillover' is negative the impacts on workers' health can be significantly damaging. As a result of this understanding, organisational OHS programmes are increasingly being broadened to incorporate strategies designed to support workers in achieving a work-life balance.

Managerial influence and safety leadership

Managerial influence

Early OHS management approaches focused upon engineering and technological solutions to emerging safety hazards. However, since the mid-twentieth century, there has been a growing interest in the impact of management and organisational factors upon OHS performance (Flin 2003). Early examples of this approach were investigations by Simonds and Shafai-Sahrai (1977) and Smith *et al.* (1978). The attention given to management and organisational factors has become so significant that Hale and Hovden (1998) have referred to it as the 'third age of safety.' Various management actions have been observed in organisations demonstrating good OHS performance. These include:

1   *Management commitment to OHS.* Early studies revealed that workers' perceptions of managers' commitment to OHS were a major factor in the success of an organisation's OHS programme (Zohar 1980). Management commitment has been referred to as a necessary condition for a safe workplace (Shannon *et al.* 2001).

2   *Worker participation in the OHS process.* Workers are recognised to be in the best position to make suggestions about OHS improvements and teams have been found to make better OHS decisions than individuals (Culvenor 2003). The encouragement of upward communication and involvement in decision-making also has an empowering effect, providing employees with authority, responsibility and accountability (Vassie and Lucas, 2001). However, while encouraging worker involvement, managers must also take care not to abrogate their managerial responsibility for OHS (Roy, 2003).

3   *Provision of OHS training.* In order to actively participate in the OHS process, it is critical that workers are provided with adequate OHS training. OHS training programmes should be carefully designed after a comprehensive assessment of the organisation's needs, i.e. what OHS knowledge, skills and abilities are missing that are required to enable employees at all levels to perform their jobs safely?

4   *Hiring practices.* Where organisations establish recruitment criteria that are designed to ensure the selection of people who are safety conscious, organisational OHS performance is reported to be better. Also, when the organisation actively strives to communicate organisational OHS values and commitment to prospective workers, it is more likely to recruit workers with compatible OHS attitudes and expectations.

5   *Reward systems.* Incentive programmes that reinforce desired OHS behaviours are a feature of organisations with good OHS performance. Vredenburgh (2002) suggests that rewards can include informational (e.g. feedback), social (e.g. praise/recognition) and tangible (e.g. bonuses/awards) reinforcement for OHS perfromance.

6   *Communication and feedback.* Clear and consistent communication of OHS expectations is vital to OHS performance. The provision of constant feedback to workers about their OHS performance is also critical because many serious OHS incidents occur as a result of actions that are routinely undertaken but which, in most instances, do not result in injury.

One striking feature of research into managerial and organisational determinants of OHS performance is the consistency with which these management actions have been linked to high levels of OHS performance

(Varonen and Mattila 2000). This consistency demonstrates that managers' actions are an extremely important factor in determining an organisation's OHS performance.

Researchers have examined the role played by different levels of management in shaping OHS performance and, consequently, there is a growing understanding of the need to pay attention to managerial actions at all levels within the organisation. Senior managers play a key role in establishing an organisation's OHS policy, setting strategic objectives for OHS and allocating organisational resources to the overall management of OHS. It is critical that managers at this level are seen to take OHS seriously, demonstrating their commitment to OHS through their actions. However, workers 'at the coalface' are likely to have little direct contact with senior management and the role played by middle managers and first level supervisory personnel is equally critical. Where middle managers or supervisors do not behave in a manner which is consistent with espoused organisational OHS policy or value statements, these policies and values are unlikely to be put into practice.

The influence of supervisors on safety performance is likely to be increased in the construction context because construction work is highly decentralised, with productive work undertaken at sites remote from the corporate office. This geographical dispersion is likely to increase the behavioural influence of supervisors relative to senior management. Construction work is also largely non-routine, necessitating the exercise of supervisory discretion in the interpretation of formal safety policies and procedures. In this context, the role of supervisors in shaping subordinates' safety behaviour is likely to be considerably greater than in stable work contexts characterised by routine production processes. The importance of first-level supervisors in the construction context is illustrated in the Case Study below.

Case Study 6.2

Supervisory safety leadership in the construction industry

A research project was undertaken to investigate the extent to which supervisors influence the injury rates of the workgroups they supervise in the Australian construction industry. Data was collected within three organisations as follows:

Study one (n=71) was undertaken within the regional construction and maintenance works district of a state-based road construction and maintenance organisation in the south east of Australia;

Study two (n=99) was undertaken at a hospital construction project in Melbourne; and

Study three (n=137) was undertaken at the Melbourne operations of a national steel reinforcement manufacturing organisation.

Data was collected using a survey tool specially developed to measure workers' perceptions of their immediate supervisors' safety response, using a survey instrument developed by Zohar (2000). Supervisors' safety response is understood to be a facet of group-level safety climate, which is linked to the safety performance of workgroups in non-construction settings (Zohar, 2000). The research also measured workers' perceptions of the organisation-level safety climate, i.e., the 'shared perceptions of the organisation's practices and policies relating to safety' (Kath *et al.* 2010, p. 1489).

Data was analysed to test whether supervisors' safety response mediated the relationship between perceptions of the organisational safety climate and the workgroup injury frequency rates within the three organisations. Analysis of the data confirmed that the relationship between perceptions of the organisational safety climate and workgroup injury frequency rate were fully mediated by perceptions of supervisors' safety response. Thus, workers' perceptions of organisational policies and practices relating to safety impact upon OHS performance indirectly, through supervisors' safety responses. The results highlight the critical role played by first-level supervisors in communicating to workers 'what the organisation really wants' in relation to OHS. This suggests a 'cascading' managerial influence by which management commitment to safety filters down through organisational hierarchies. Supervisors act as a 'conduit' through which organisational safety priorities are communicated and provide important feedback to front-line workers concerning the appropriateness of their behaviour (Lingard *et al.* 2009).

The research also sought to examine the mechanisms of influence among subcontracted workgroups at the hospital construction project. Workers' perceptions of the safety policy and practices of the principal contractor positively and significantly predicted subcontracted workers' perceptions of their own organisations' safety policy and practices. Thus, when subcontracted workers perceived the principal contractor was serious about OHS, they were more likely to perceive that their direct employer (i.e., the subcontractor who employs them organisation) is similarly committed to OHS.

First-level supervisors (foremen and leading hands) played a critical role in shaping workgroup safety performance. Perceptions of the subcontractor supervisors' safety response were inversely related to subcontracted workgroup's injury rate. It is through supervisors thatthe behavioural benefits of organisational safety policies and practices are realised. When supervisors establish high expectations for safety, are consistent in their actions and engage in frequent safety-related interactions with workers, safety performance is likely to be enhanced.

The research highlights the role played by supervisors as a critical conduit through which organisation's safety values and priorities are communicated to workers. Supervisory personnel are particularly influential because they 'filter' organisational OHS messages and shape employees' beliefs about the importance of OHS relative to other organisational goals, e.g. production and cost.

Several researchers have tried to modify supervisors' safety leadership behaviour to improve workgroup OHS performance (Zohar, 2002). More recently, Kines *et al.* (2010) examined supervisors' interactions with workers in the Danish construction industry. They evaluated a 'safety coaching' intervention in which supervisors (foremen) were encouraged to discuss safety with workers and provide proactive and supportive feedback about safe ways of working.

Baseline measurements revealed that foremen spoke with their workers several times a day but that safety was only mentioned in between 6 and 16 per cent of these verbal exchanges.

Supervisors who were coached and given bi-weekly feedback increased the safety content of their interactions with workers, relative to supervisors who received no coaching or feedback. Also, the levels of safety observed in workgroups whose supervisors received coaching and feedback improved significantly over baseline conditions. This suggests that the OHS performance of workers can be positively influenced by interventions designed to improve the safety management practices of first-level supervisors in the construction industry.

Trust and worker involvement

Recent research has focused on the social context of work and the role played by trust and mutual understanding, particularly between occupants of different levels within organisational hierarchies, in influencing OHS outcomes (Clarke, 2006). Conchie and Donald define trust as 'an individual's willingness to rely on another (i.e. a leader) based upon positive expectations that he or she will act safely or intend to act safely' (p. 137). Kath *et al.* (2010) present evidence indicating that a consistent approach to the application of organisational safety policies and practices demonstrates a concern for workers' health and safety and helps to establish trust. In an analysis of safety leadership in the UK construction industry, Conchie and Donald (2009) reveal that the positive impact of transformational safety leadership is contingent upon the extent to which workers perceive their supervisors as trustworthy in relation to OHS. In conditions of low trust, transformational leadership has no significant impact upon workers' safety citizenship behaviour. McLain and Jarrell (2007) suggest that trust encourages the sharing and acceptance of safety information, which increases knowledge about how to integrate task and safety demands and reduces uncertainty about what is expected. Where workers trust managers' handling of OHS, they are more likely to believe that safety and production targets are compatible (McLain and Jarrell, 2007). Luria (2010) found that trust between leaders and subordinates was directly linked to reduced injury rates within workgroups. Thus it seems that high-trust relationships between supervisors and workers create a safer work environment. Also, over time, workers who perceive that their supervisors place importance upon OHS will develop strong and positive work group safety climates, i.e. a perception that safety is a priority for their workgroup. Not only does trust seem to be an important determinant of safety within organisations but research shows that it is also a predictor of effective rehabilitation and return-to-work of injured workers (Baril *et al.* 2003).

Given the importance of trust to workers' health, safety and wellbeing, there may be a need to re-think traditional management approaches to OHS. Traditionally OHS management has assumed workers are motivated to exert only enough effort as is necessary for task completion. Management approaches to OHS have been based upon rules to enforce behaviour and relied upon managers' use of legitimate authority to control workers' behaviour (Zacharatos *et al.* 2005). Zacharatos *et al.* (2005) suggest that workers' commitment to achieving organisational goals can be enhanced by encouraging participative decision-making, treating workers with respect, sharing information and providing high-quality training. This shift from a 'control-based' to a 'social exchange' approach to the management of OHS hinges upon the workers' belief in the trustworthiness of management with regard to workers' OHS. Trust is enhanced in situations in which there is open communication between workers and managers and workers' suggestions for safety improvements are sought and taken seriously. Törner and Poussette (2009) report that Swedish construction organisations with high safety standards adopt joint problem-solving and involve workers and their representatives in OHS decision-making. Worker participation in the development and implementation of organisational OHS initiatives can facilitate the identification of practical and sustainable solutions that are more likely to yield long-term OHS improvements (Rasmussen *et al.* 2006).

Conclusions

Workers' health, safety and wellbeing are critical to the construction industry's sustainability and long-term economic performance. Yet the industry faces many difficult challenges. The International Labour Organization (ILO) estimated at least 60,000 fatal accidents occur each year on construction sites around the world – representing one in six of all fatal work-related accidents (ILO, 2005). In developed economies the construction industry is characterised by an intransigent residual level of workers' death, injury and illness that seems resistant to change. In developing countries, the situation is much worse. For example, according to Wells and Hawkins (no date) there are, on average, 100 fatalities on UK construction sites each year. However, the ILO estimated the number of deaths from accidents at work in India in 2001 to be 40,000, with an additional 262,000 workers dying from work-related diseases and 65,000 from dangerous substances. Assuming that construction accounts for one sixth of these deaths, then over 60,000 Indian workers died as a result of work exposure in a single year (Wells and Hawkins, no date).

It is often argued that 'safety saves money' and therefore there is a natural cost incentive for organisations to invest in OHS. Although the aggregated costs of work death, injury and illness are very substantial, Australian data suggests that the largest proportion of these costs is borne by the community (e.g. in social security and welfare payments) or by injured workers and their families (e.g. in the provision of care for injured or ill persons). The proportion of costs of work-related death injury and illness borne by employers is relatively small. Further, this is only one side of the OHS equation. In order to determine whether there is a cost incentive for construction firms to prevent deaths, injuries and illnesses, it is necessary to compare the costs of harm with the costs of prevention. At the level of a single enterprise, e.g. a construction firm, profitability is usually measured on an annual basis. Given this accounting practice, it is unlikely that the return on investment in a period of one year would be sufficient to justify significant investment in preventive measures, particularly in the case of high severity but low frequency events.

Given that this is the case, the regulation of OHS and other aspects impacting workers' wellbeing, such as work hours, remains critical and governments have an important role to play in ensuring that minimum standards are maintained.

Ensuring equity and protecting the health, safety and wellbeing of vulnerable worker groups is a growing challenge. There is considerable evidence that young workers, migrant workers and workers whose employment arrangements are precarious are high risk groups for work-related harm. Further, it is these groups who may find it difficult to engage in organised labour activities, participate in worker–employer consultative processes and access the support and advice they need to equip them to advocate for improvements in work conditions.

Long work hours and fatigue remain a problem for construction workers' health, safety and wellbeing. The problems associated with long hours and fatigue are experienced by both blue and white collar workers, and are particularly acute in project-based work. In the author's research on one large construction project in Melbourne, a managerial employee explained the issue of burnout related to long hours as follows:

> We burn a lot of people out who would otherwise stay in the industry. I've seen projects where we've had engineers covering two shifts working 16 hours and they've got an hour drive to and from work and they end up having six hours at home and they do that for nine months and they give it away, not only the project away, they give the industry away and we just burn them out

It is essential that construction organisations consider the impact of work schedules on workers' health, safety and wellbeing and ensure that sufficient opportunities for rest and recovery are provided to mitigate some of these problems.

Much work has been done in the analysis of causes of work-related injury and we now have robust models to explain how injuries occur. It is fair to say that, in general, we know what causes work-injuries and by logical extension, we know what needs to be addressed to prevent similar injury events from happening in the future. There is also a growing understanding of the importance of the social context of work and we are developing a good understanding of the importance of safety leadership and trust between workers and managers in ensuring a safe and healthy work environment. It is likely that to achieve sustainable improvements in OHS in the construction industry, organisations will need to move away from the traditional control-based approach to managing human resources and adopt the more participatory, high-involvement approach. Unfortunately, this type of change will not be easy to implement and would necessarily challenge prevailing neo-liberal concepts of managerial prerogative.

References

Allen, T. D. and Armstrong, J. (2006) Further examination of the link between work–family conflict and physical health: The role of health-related behaviours, *American Behavioral Scientist*, 49, 1204–21.

American Institute for Cancer Research (2007) *Food, Nutrition and the Prevention of Cancer: A Global Perspective*, Washington DC, American Institute for Cancer Research.

Arditi, D. and Chotibhongs, R. (2005) Issues in subcontracting practice, *Journal of Construction Engineering and Management*, 131, 866–76.

Australian Bureau of Statistics, (2008) 4837.0.55.001, *Health of Mature Age Workers in Australia: A Snapshot, 2004–05*, Canberra.

Baril, R., Clarke, J., Friesen, M., Stock, S., Cole, D. the Work Ready Group (2003) Management of return-to-work programs for workers with musculoskeletal disorders: a qualitative study in three Canadian provinces, *Social Science and Medicine*, 57, 2101–14.

Barnes, C. M. and Wagner, D. T. (2009) Changing to daylight saving time cuts into sleep and increases workplace injuries, *Journal of Applied Psychology*, 94, 1305–17.

Bluff, E. and Gunningham, N. (2003) *Principle, Process, Performance or What? New Approaches to OHS Standards Setting,* Working Paper 9, National Research Centre for Occupational Health and Safety Regulation, Australian National University, Canberra, online at www.ohs.anu.edu.au.

Bongers, P. M., Kremer, A. M. and ter Laak, J., (2002) Are psychosocial factors, risk factors for symptoms and signs of the shoulder, elbow, or hand/wrist?: A review of the epidemiological literature, *American Journal of Industrial Medicine*, 41, 315–342.

Breslin, F. C., Polzer, J., MacEachen, E., Morrongiello, B. and Shannon, H. (2007) Workplace injury or 'part of the job'?: Towards a gendered understanding of injuries and complaints among young workers, *Social Science and Medicine*, 64, 782–93.

Burt, C. D. B., Chmiel, N. and Hayes, P. (2009) Implications of turnover and trust for safety attitudes and behaviour in work teams, *Safety Science*, 47, 1002–06.

Bust, P. D., Gibb, A. G. F. and Pink, S. (2008) Managing construction health and safety: Migrant workers and communicating safety messages, *Safety Science*, 46, 585–602.

Caruso, C. C., Bushnell, T., Eggerth, D., Heitmann, A., Kojola, B., Newman, K., Rosa, R. R., Sauter, S. L. and Vila, B. (2006) Long working hours, safety and health: Toward a national research agenda, *American Journal of Industrial Medicine*, 49, 930–42.

Chang, F-L., Sun, Y-M., Chuang, K-H. and Hseu, D-J. (2009) Work fatigue and physiological symptoms in different occupations of high-elevation construction work, *Applied Ergonomics*, 40, 591–6.

Chin, P., DeLuca, C., Potyh, C., Chadwick, I., Hutchinson, N. and Munby, H. (2010) Enabling youth to advocate for workplace safety, *Safety Science*, 48, 570–9.

Clarke, S., (1999) Perceptions of organizational safety: implications for the development of safety culture, *Journal of Organizational Behavior*, 20, 185–98.

Conchie, S. M. and Donald, I. J., (2009) The moderating role of safety-specific trust on the relation between safety-specific leadership and safety citizenship behaviors, *Journal of Occupational Health Psychology*, 14, 137–47.

Culvenor, J. (2003) Comparison of team and individual judgments of solutions to safety problems, *Safety Science*, 41, 543–56.

Dawson, S., Willman, P., Clinton, A, and Bamford, M. (1988) *Safety at Work: The limits of self-regulation*, Cambridge University Press, Cambridge.

DeGroot, T. and Kiker, D. S. (2003) A meta-analysis of the non-monetary effects of employee health management programmes, *Human Resource Management*, 42, 53–69.

Dembe, A. E., Erickson, J. B., Delbos, R. G. and Banks, S. M. (2005) The impact of overtime and long work hours on occupational injuries and illnesses: New evidence from the United States, *Occupational and Environmental Medicine*, 62, 588–97.

Devine, C. M., Jastran, M., Jabs, J., Wethington, E., Farell, T. J. and Bisogni, C. A. (2006), A lot of sacrifices: Work–family spillover and the food choice coping strategies of low-wage employed parents, *Social Science and Medicine*, 63, 591–603.

Djebarni, R. (1996) The impact of stress in site management effectiveness, *Construction Management and Economics*, 14, 281–93.

Dong, X. S., Fujimoto, A., Ringen, K. and Men, Y. (2009) Fatal falls among Hispanic construction workers, *Accident Analysis and Prevention*, 41, 1047–52.

Drennan, F. S., Ramsay J. D. and Richey, D. (2006) Integrating employee safety and fitness: A model for meeting NIOSH's Steps to a Healthier US Workforce challenge, *Professional Safety*, 51, 26–35.

Driscoll, T. (2006) Fatal injury of young workers in Australia, *Journal of Occupational Health and Safety: Australia and New Zealand*, 22, 151–61.

Emberland, J. S. and Rundmo, T. (2010) Implications of job insecurity perceptions and job insecurity responses for psychological wellbeing, turnover intentions and reported risk behaviour, *Safety Science*, 48, 452–9.

Fabiano, B., Currò, F., Reverberi, A. P. and Pastorino, R. (2008) A statistical study on temporary work and occupational accidents: Specific risk factors and risk management strategies, *Safety Science*, 46, 535–44.

Flin, R., (2003) 'Danger – men at work': Management influence on safety, *Human Factors and Ergonomics in Manufacturing*, 13, 261–8.

Folkard, S. and Lombardi, D. A., (2006) Modeling the impact of components of long work hours on injuries and 'accidents', *American Journal of Industrial Medicine*, 49, 953–63.

Franche, R. L., Williams, A., Ibrahim, S., Grace, S. L., Mustard, C., Minore, B. and Stewart, D. E. (2006). Path analysis of work conditions and work–family spillover as modifiable workplace factors associated with depressive symptomatology, *Stress and Health*, 22, 91–103.

Gallagher, C., Underhill, E and Rimmer, M. (2003) Occupational safety and health management systems in Australia: barriers to success, *Policy and Practice in Health and Safety*, 2, 67–81.

Government of South Australia, (2009) *South Australian Construction Industry Workforce Action Plan (2009 – 2014)*, Adelaide.

Greenwood, M. R. (2002) Ethics and HRM: A review and conceptual analysis, *Journal of Business Ethics*, 36, 261–78.

Guadalupe, M. (2003) The hidden costs of fixed term contracts: the impact on work accidents, *Labour Economics*, 10, 339–57.

Guberon, E. and Usel, M. (1998) Permanent work incapacity, mortality and survival without incapacity among occupations and social classes: a cohort study of aging men in Geneva, *International Journal of Epidemiology*, 27, 1026–32.

Hale, A. R. and Hovden, J. (1998) Management and culture: The third age of safety. A review of approaches to organizational aspects of safety, health and environment, A. M. Feyer and Williamson A. (eds), *Occupational Injury: Risk prevention and intervention*, 129–65, Taylor & Francis, London.

Haslam, R. A., Hide, S. A., Gibb, A. G. F., Gyi, D. E., Pavitt, T., Atkinson, S. and Duff, A. R. (2005) Contributing factors in construction accidents, *Applied ergonomics*, 6, 401–415.

Haynes, N. S. and Love, P. E. D. (2004) Psychological adjustment and coping among construction project managers, *Construction Management and Economics*, 22, 129–40.

Heinrich H. W. (1959) *Industrial accident prevention* (4th edn). McGraw-Hill: New York.

Hillier, D., Fewell, F., Cann, W. and Shephard, V. (2005) Wellness at work: Enhancing the quality of our working lives, *International Review of Psychiatry*, 17, 419–31.

Hobman, E. V., Jimmieson, N. L. and McDonald, K. (2010) *An Examination of Fatigue in the Construction Industry*, The Centre for Organizational Psychology, University of Queensland, Brisbane, February 2010.

Hopkins, A. (1999) For whom does safety pay? The case of major accidents, *Safety Science*, 32, 143–53.

HSE (Health and Safety Executive) (2003) *Causal Factors in Construction Accidents*, Research Report 156, HMSO Books, Norwich.

International Labour Organization, (2005) Estimate by ILO on the occasion of the World Day for Safety and health at Work, 2005 www.ilo.org/global/About_the_ ILO/Media_and_public_ information/Press_releases/lang-en/WCMS_ 005161/ index.htm.

Jamal, M. (2004) Burnout, stress and health of employees on non-standard work schedules: a study of Canadian workers, *Stress and Health*, 20, 113–19.

Janicak, C. A. (2008) Occupational fatalities due to electrocutions in the construction industry, *Journal of Safety Research*, 39, 617–21.

Jansen, N. W. H, Kant, I. J., von Amelsvoort, L. G. P. M., Kristensen, T. S., Swaen, G. M. H., and Nijhuis, F. J. N. (2006) Work-family conflict as a risk factor for sickness absence, *Occupational and environmental medicine*, 63, 488–494.

Kath, L. M., Magley, V. J. and Marmet, M. (2010) The role of organizational trust in safety climate's influence on organisational outcomes, *Accident Analysis and Prevention*, 42, 1488–97.

Katsakiori, P., Sakellaropoulos, G. and Manaatakis, E. (2009) Towards and evaluation of accident investigation methods in terms of their alignment with accident causation models, *Safety Science*, 47, 1007–15.

Kines, P., Anderson, L. P. S., Spangenberg, S., Mikkelsen, K. L., Dyreborg, J. and Zohar, D. (2010) Improving construction site safety through leader-based verbal safety communication, Journal of Safety Research, 41, 399–406.

Korda, R. J., Strazdins, L., Broom, D. H. and Lim, L., L-Y. (2002) The health of the Australian workforce: 1998–2001, *Australian and New Zealand Journal of Public Health*, 26, 325–31.

Larsson, T. and Field, B. (2002) The distribution of occupational risks in the Victorian construction industry, *Safety Science*, 40, 439–56.

Lavack, A. M., Magnuson, S. L., Deshpande, S., Basil, D. Z., Basil, M. D. and Mintz, J. H. (2008) Enhancing occupational health and safety in young workers: the role of social marketing, *International Journal of Nonprofit and Voluntary Sector Marketing*, 13, 193–204.

Leung, M-Y., Chan, Y. S. and Olomolaiye, P. (2008) Impact of stress on the performance of construction project managers, ASCE *Journal of Construction Engineering and Management,* 134, 644–52.

Leung, M-Y., Chan, Y. S. and Yuen, K. W. (2010) Impacts of stressors and stress on the injury incidents of construction workers in Hong Kong, ASCE *Journal of Construction Engineering and Management*, 136, 1093–103.

Leung, M-Y., Chan, Y. S. and Yu, J. (2008) Integrated model for the stressors and stresses of construction project managers in Hong Kong, ASCE *Journal of Construction Engineering and Management,* 134, 126–34.

Leung, M-Y., Skitmore, M. and Chan, Y. S. (2007) Subjective and objective stress in construction cost estimation, *Construction Management and Economics*, 25, 1063–75.

Lingard, H. and Francis, V. (2004) A comparative study of the work–life experiences of men and women working in office and site-based roles in the Australian construction industry, *Construction Management and Economics*, 22, 991–1002.

Lingard, H. and Francis, V. (2005) Does work–family conflict mediate the relationship between job schedule demands and burnout in male construction professionals and managers?, *Construction Management and Economics*, 23, 733–45.

Lingard, H., Cooke, T. and Blismas, N. (2009) Group-level safety climate in the Australian construction industry: Within-group homogeneity and between-group differences in road construction and maintenance, *Construction Management and Economics*, 27, 419–32.

Lingard, H., Cooke, T. and Blismas, N. (2010) Safety climate in conditions of construction subcontracting: A multi-level analysis, *Construction Management and Economics*, 28, 313–825.

Linton, S. J., (2001) Occupational psychological factors increase the risk for back pain: A Systematic Review, *Journal of Occupational Rehabilitation*, 11, 5–66.

Loosemore, M. and Andonakis, N. (2007) Barriers to implementing OHS reforms – The experiences of small subcontractors in the Australian Construction Industry, *International Journal of Project Management*, 25, 579–88.

Loughlin, C. and Frone, M. R. (2004) Young workers' occupational safety, *The Psychology of Workplace Safety*, Barling, J. and Frone M. (eds), American Psychological Association: Washington DC, 107–25.

Love, P. E. D., Edwards, D. J. and Irani, Z. (2010) Work stress, support, and mental health in construction, ASCE *Journal of Construction Engineering and Management*, 136, 650–58.

Lundberg, J., Rollenhagen, C. and Hollnagel, E. (2009) What-you-look-for-is-what-you-find – The consequences of underlying accident causation models in eight accident investigation manuals, *Safety Science*, 47, 1297–311.

Luria, G. (2010) The social aspects of safety management: Trust and safety climate, *Accident Analysis and Prevention*, 42, 1288–95.

McLain, D. L. and Jarrell, K. A., (2007) Perceived compatibility of safety and production expectations in hazardous occupations, *Journal of Safety Research*, 38, 299–309.

Manu, P., Ankrah, N., Proverbs, D. and Suresh, S. (2010) An approach for determining the extent of contribution of construction project features to accident causation, *Safety Science*, 48, 667–92.

Maslach, C., Schaufeli, W. B. and Leiter, M. P. (2001) Job burnout, *Annual Review of Psychology*, 52, 397–422.

Mayhew, C., Quinlan, M. and Ferris, R. (1997) The effects of subcontracting/outsourcing on occupational health and safety: Survey evidence from four Australian industries, *Safety Science*, 25, 163–78.

Miller, P. and Haslam, C. (2009) Why employers spend money on employee health: Interviews with occupational health and safety professionals from British Industry, *Safety Science*, 47, 163–9.

National Occupational Health and Safety Commission (2004) *The costs of work-related injury and illness for Australian employers, workers and the community*, Commonwealth Government of Australia, Canberra.

Ng, T. S., Skitmore, M. R. and Leung, T. K. C. (2005) Manageability of stress among construction project participants, *Engineering, Construction and Architectural Management*, 12, 264–82.

Nichols, T. (1997) *The Sociology of Industrial Injury*, Mansell Publishing, London.

Olds, D. M. and Clarke, S. P. (2010) The effect of work hours on adverse events and errors in health care, *Journal of Safety Research*, 41, 153–62.

Peregrin, T. (2005) Weighing in on corporate wellness programs and their impact on obesity, *Journal of the American Dietetic Association*, 1192–4.

Quinlan, M. (2007) Organizational restructuring/downsizing, OHS regulation and worker health and wellbeing, *International Journal of Law and Psychiatry*, 30, 385–99.

Rasmussen, K., Glasscock, D. J., Hansen, O. N., Carstensen, O., Jepsen, J. F. and Nielsen, K. J. (2006) Worker participation in change processes in a Danish industrial setting, *American Journal of Industrial Medicine*, 49, 767–79.

Rowan, J. J. (2000) The moral foundation of employee rights, *Journal of Business Ethics*, 25, 355–61.

Roy, M., (2003) Self-directed workteams and safety: a winning combination? *Safety Science*, 41, 359–76.

Sang, K. J. C., Dainty, A. R. J. and Ison, S. G. (2007) Gender: a risk factor for occupational stress in the architectural profession?, *Construction Management and Economics*, 25, 1305–17.

Simonds, R. H. and Shafai-Sahrai, Y. (1977) Factors apparently affecting injury frequency in eleven matched pairs of companies, *Journal of Safety Research*, 9, 120–27.

Smith, J. J., Cohen, H. H., Cohen, A. and Cleveland, R. (1978) Characteristics of successful safety programs, *Journal of Safety Research*, 10, 5–15.

Sonnentag, S. and Zijlstra, F. R. H. (2006) Job characteristics and off-job activities as predictors of need for recovery, wellbeing and fatigue, *Journal of Applied Psychology*, 91, 330–50.

Sorensen, G., Sembajwe, G., Harley, A. and Quintiliani, L. (2009) Work and occupation: Important indicators of socioeconomic position and life experiences influencing cancer disparities, Koh, H. K. (ed), *Toward the Elimination of Cancer Disparities*, Springer, New York, 83–105.

Suraji, A., Duff, A. R. and Peckitt, S. J. (2001) Development of causal model of construction accident causation, *Journal of Construction Engineering and Management*, 127, 337–44.

Sutherland, V. J. and Davidson, M. J. (1989) Stress among construction site managers: a preliminary study, *Stress Medicine*, 5, 221–35.

Törner, M. and Poussette, A., (2009) Safety in construction – a comprehensive description of the characteristics of high safety standards in construction work, from the combined perspective of supervisors and experienced workers, *Journal of Safety Research*, 40, 399–409.

Trajkovski, S. and Loosemore, M. (2006) Safety implications of low-English proficiency among migrant construction site operative, *International Journal of Project Management*, 24, 446–52.

Tucker, S., Chmiel, N., Turner, N., Hershcovis, M. S. and Stride, C. B. (2008) Perceived organizational support for safety and employee safety voice: The mediating role of co-worker support for safety, *Journal of Occupational Health Psychology*, 13, 319–30.

Vassie, L. H. and Lucas, W. R. (2001) An assessment of health and safety management within working groups in the UK manufacturing sector, *Journal of Safety Research*, 32, 479–90.

Vredenburgh, A. G. (2002) Organizational safety: Which management practices are most effective in reducing employee injury rates, *Journal of Safety Research*, 33, 259–76.

Wang, J. L., Afifi, T. O., Cox, B. and Sareen, J. (2007). Work–family conflict and mental disorders in the United States: Cross-sectional findings from the National Comorbidity Survey, *American Journal of Industrial Medicine*, 50, 143–9.

Wells, J. and Hawkins, J. (no date) *Promoting Construction Health and Safety through Procurement: A briefing note for developing countries,* Institution of Civil Engineers, London, www.engineersagainstpoverty.org/_db/_documents/EAP-ICE_HS_Briefing_Note.pdf.

Williams, Q. Jr., Ochsner, M., Marshall, E., Kimmel, E. and Martino, C. (2010) The impact of a peer-led participatory health and safety training program for Latino day laborers in construction, *Journal of Safety Research*, 41, 253–61.

Williamson, A., Lombardi, D. A., Folkard, S., Stutts, J., Courtney, T. K. and Connor, J. L. (2011) The link between fatigue and safety, *Accident Analysis and Prevention*, 43, 498–515.

Zacharatos, A., Barling, J. and Iverson, R. D., (2005) High-performance work systems and occupational safety, *Journal of Applied Psychology*, 90, 77–93.

Zohar, D. (2000) A group-level model of safety climate: testing the effect of group climate on micro-accidents in manufacturing jobs, *Journal of Applied Psychology*, 85, 587–96.

Zohar, D. (2002) Modifying supervisory practices to improve subunit safety: a leadership-based intervention model, *Journal of Applied Psychology*, 87, 156–63.

7 Equality, diversity, inclusion and work–life balance in construction

Katherine Sang and Abigail Powell

Introduction

Despite a range of targeted initiatives the profile of the construction industry workforce has shown little, if any, change over the last 20 years. It seems that the sector's workforce is still largely white, male and able-bodied. This situation means that the industry is failing to exploit the potential pool of recruits and as such may be missing out on highly skilled and committed employees. This chapter brings together a range of international literature from construction, and human resource management (HRM) more generally, to address the importance of equality, diversity, inclusion (EDI) and work–life balance (WLB) issues for the construction sector, drawing upon both employer and employee perspectives. It also provides a critique of current practices within the industry.

We begin by defining our key concepts: equality, diversity, inclusion, work–life balance and organizational culture. We also outline the main arguments for EDI and WLB from an ethical perspective, founded in a commitment to social justice for all and a business perspective, founded in the belief that these concepts can improve efficiency and economic growth. Following this, we provide an overview of the situation of women, ethnic minorities, migrant workers, people with disabilities and older workers in the construction industry, as well as work–life balance in the construction sector. Finally, the chapter outlines some of the cultural and structural factors that have hindered progression in construction in terms of EDI and WLB, before making recommendations to address these issues.

Equality, diversity and inclusion

Equality is 'a state or condition of being the same, especially in terms of social status or legal/political rights' (Pilcher & Whelehan, 2004). Movements towards equality therefore often focus on minority groups, such as women, being treated the same as men, irrespective of their differences. However, this concept has been problematised for failing to recognise differences between and within minority and majority groups, and for implying that the goal of equality can be achieved if subordinated groups assimilate to the culture of the dominant group (Pilcher & Whelehan, 2004). In an attempt to address deficiencies with the concept of 'equality', theorists have more recently proposed the term 'diversity'.

Savoie and Sheehan (2001:2) state that 'diversity includes all characteristics and experiences that define each of us as individuals ... a common misconception about diversity is that only individuals or groups with particular attributes are included under its umbrella. Exactly the opposite is true. Diversity encompasses the entire spectrum of primary attributes of individuals including: Race, Ethnicity, Gender, Age, Religion, Disability and Sexual Orientation'.

Kossek and Pichler (2009) suggest that workforce diversity is a variation of social and cultural identities among people working in a defined employment setting. However, an organisation can be numerically diverse without being inclusive or multicultural. Loosemore *et al.* (2003) suggest that the concept of diversity recognises that there are differences between people, which if harnessed can create a more productive, adaptable and creative work environment, where people feel values and their talents are fully utilised. They argue that organisational diversity is usually internally initiated from within the organisations, is driven by business needs, is focused on improving the work environment, concentrates on proactively taking advantage of opportunities, and assumes pluralism.

The Equalities Review (2007:18) uses a definition that couples equal opportunity with diversity: 'an equal society protects and promotes equal, real freedom and substantive opportunity to live in the ways people value and would choose, so that everyone can flourish.' Bagilhole (2009) suggests that this definition recognises people's different needs, situations and goals and removes the barriers that limit what people can do and can be. It recognises that some people may need more and different resources to enjoy genuine freedom and fair access to opportunities, in different ways and at different times; 'it recognises that equality does not mean sameness or same treatment' (2009:53).

An inclusive workplace is one that values individual differences (Kossek & Pichler, 2009), but does not necessarily treat everyone the same. Kossek and Pichler (2009) suggest that the current trend in fostering diversity effectiveness within organisations is developing equal opportunities strategies that foster an 'inclusive workplace'.

Work–life balance

One way organisations can help foster inclusive workplaces is by adopting policies that enable employees to achieve work–life balance (WLB). WLB is about individuals balancing their home and work lives, so that they are able to enjoy both without either negatively impacting on the other. WLB policies are often discussed interchangeably with family-friendly policies. In the UK, national legislation in this area includes employee access to maternity leave, maternity pay, the right of mothers to return to work, paternity leave, parental leave and the right of some employed parents and carers to request flexible working arrangements.[1]

However, WLB policies have been criticised for adopting a 'woman's problem' approach to equality (Liff & Cameron, 1997), primarily because they fail to recognise and challenge the gendered division of paid work, domestic work and childcare (Bagilhole, 2009). When WLB policies are directed primarily at women they can actually perpetuate gender inequities (Lewis & Roper, 2008) by marginalising those who use them. This is a cyclical process given that evidence of marginalisation is likely to further discourage others (such as men) from taking them up, thereby reinforcing the idea that only non-standard workers, not fully committed to work or career will use these policies. In addition, much WLB research focuses on work–to–family conflict. It is important to recognise that the relationship can operate in reverse and that non-caring activities, for example, social activities are also an important aspect of good WLB.

Culture

An organisation's culture is particularly important in terms of the extent to which minority groups and equality and diversity policies are accepted by employees, and to which individuals feel able to take-up WLB policies without facing disadvantage. We argue that understanding the experiences of those who are 'outsiders' in construction, for example, women, ethnic minorities and people with disabilities, can help to make visible the dominant culture within the sector which often remains invisible or taken for granted (Kronsell, 2005; Sang *et al.*, 2010a). Within construction, there is likely a particular industry culture as well as sub-cultures within different companies and locations (Greed, 2000).

While we do not attempt to describe construction culture, it is important to explain what culture is, in order to understand how it may impact on the industry, individuals and policies. Culture is the 'view of the world' shared by members of an organisation and guiding their behaviour (Jex & Britt, 2008). It is the 'shared mental assumptions that guide interpretation and actions in organizations by defining appropriate behaviour for various situations' (Ravisi and Schultz, 2006: 437). For more information on organisational culture see Alasuutari, 1995; McIlwee and Robinson, 1992; Greenwood, 1997; Wajcman, 1998; Brown, 1995; Janson, 1994; Schein, 1992. While construction organisations may have policies on EDI and WLB, an organisation's culture will impact significantly on how these policies are put into practice (e.g., Jex & Britt, 2008). More generally, researchers have also argued that culture is gendered and that in workplaces such as the construction industry, traditional masculinities are a dominant element of the culture (Hofstede, 1984; see also Hofstede, 2003; Mills, 1988; Bottero, 1992; Itzin, 1995).

Arguments for equality, diversity, inclusion, work–life balance

The business case

The use of the business case is now widespread and suggests there are a range of economic benefits resulting from diverse workforces. A more diverse workforce may, for example, work more effectively (Ely & Thomas, 2001), widen the available pool of talent, reduce staff turnover and attrition, enhance innovation and creativity and improve the image and reputation of the industry (EHRC, 2009). Experiences of other sectors suggest that improving opportunities for minority groups promotes a more informed workforce and organisations that are more adaptable, closer to their customers (Greenhaus & Callanan, 1994), and more responsive to market changes (Coussey & Jackson, 1991). For example, robust HRM processes combined with fairness in the promotion of women have been associated with enhanced organisational performance (Harel *et al.*, 2003). Within this context fits the HRM concept of *managing diversity*. This approach promotes valuing differences between employees in order to fully exploit the potential of each employee (Torrington *et al.*, 2008). However, within the broader management literature, diversity management has been criticised for failing to address deeply embedded inequalities and power relations (Sharp *et al.*, 2011). The implication here is that such policies are implemented to achieve compliance, rather than to promote genuine cultural change (e.g., Loosemore and Phau, 2010).

Within construction, Loosemore *et al.* (2003:11) argue that providing equality of opportunity for all should form the cornerstone of good employment practice. Discrimination leads to the under-utilisation of skills and talents (Bagilhole, 1997) and stifles workforce diversity which could promote innovation and improved working (CIB, 1996).

However, there is also a wealth of literature documenting negative outcomes associated with the inappropriate management of workforce diversity (Kulik & Roberson, 2008). Diverse organisations have, for example, been associated with reduced employee commitment (Tsui *et al.*, 1992), higher employee dissatisfaction and turnover (Jackson, *et al.*, 1991) and greater intergroup conflict (Jehn *et al.*, 1999; see also EHRC, 2009). Kochan *et al.* (2003) also suggest that organisations are rarely able to capitalise on the potential benefits of workforce diversity. Nevertheless, diversity can lead to business improvements (O'Reilly *et al.*, 1998) but only if it is effectively managed (Kulik & Roberson, 2008).

Another major problem with the business case is that during an economic downtown equality, diversity and inclusion (EDI) priorities might be lowered from the organisational agenda or disappear altogether. It has been suggested that the current economic recession in the UK may result in pregnant women facing a greater likelihood of losing their job as a result of this pregnancy (Fawcett Society, 2009). As a consequence, economic arguments could in fact be used to argue against EDI (Pringle, 2009). Recent data from the UK shows that construction sector output contracted by around 13 per cent in 2009, with employment levels falling by approximately 375,000 workers between 2008 and 2010 (ConstructionSkills, 2010), making it difficult to argue for equality on the grounds of skills shortages. Therefore, while the business case has made some progress in changing employer perceptions, its use alone is problematic. Further work is needed to understand the impact of the economic crisis on equality and diversity.

In summary, the drive for diversity in the construction workforce may, at times, be opportunistic and relate to lack of appropriate skills rather than being motivated by a genuine concern for equity.

The ethical case

While the business case has some merits, over-emphasis on this perspective fails to uphold the ideal of social justice, equality and inclusivity for all (Noon, 2007). In particular, the business case implies that 'women are perhaps the 'last resort' – a suggestion that if some other source were available, WISE[2] would not be needed' (Henwood, 1996:200). It is argued, for example, that the skills shortage, rather than the development of an inclusive approach, has led more women into science, engineering and technology professions (Devine, 1992; Fielding & Glover, 1999). Importantly, this places the onus on minority groups to fit into the existing, hegemonic cultures of organisations that are employing them out of necessity, rather than a real desire for change. It is also argued that it is problematic to sell diversity discourses to the business community, which can leave women vulnerable to political and socio-economic shifts (Griffiths *et al.*, 2006).

Etzkowitz *et al.* (2000), for example, suggests that barriers to entry in both industry and academia are most readily removed in periods of economic prosperity and expansion, and prove more difficult to shift in times of recession. They cite the US, Finland and Portugal as countries where the proportion of women in research and development positions increased post-World War II. However, during periods of increased competition, discriminatory attitudes and behaviours can re-surface. The *Financial Times* (Mychasuk, 2008), reporting on feedback from a number of major European employers, found that in the current recession, employers will no longer be able to afford the 'luxury' of pursuing diversity strategies, the implication being that EDI does not make financial sense. Hogarth *et al.* (2009) found that the number of equal pay and sex discrimination claims has fallen sharply since the beginning of the recession, as a result of fears around job security. Further, the Government Equalities Office (2009) reports that nearly one-in-four men believe that it makes more sense for people on maternity leave to be made redundant first in difficult economic times. This is in contrast to a recent UK government report which emphasised the business case of employing women, particularly at senior levels (Abersoch, 2011).

Regulation and governance

The construction sector is regulated by national legislation that applies to other industries. We summarise the issues which employers and human resource managers in the construction industry must consider, before examining drivers for EDI within construction. In the UK, for example, it is unlawful to discriminate in employment, training or education on the grounds of disability, colour, race, nationality, sex, marital status or age (for further information on the relevant legislation see Bagilhole, 2009; Fredman, 2002; McColgan, 2005). Importantly, however, there is no positive duty to promote equality of opportunity and policy initiatives have not always had a direct impact on equality outcomes. However, in the UK, public sector employers have, since the Equality Act 2006, had a legal duty to pay 'due regard' to the need to take action on race, disability and gender equality. This includes undertaking equality impact assessments to evaluate the impact of new policies in terms of gender, race and disability.[3] In October 2010 a new equality act came into being in the United Kingdom. Discrimination which occurred before 1 October 2010 is subject to previous equality legislation, such as the Sex Discrimination Act (1975), The Equal Pay Act (1970), the Race Relations Act (1976) and the Disability Discrimination Act (1995). Under the Equality Act 2010, it is unlawful (except in very specific situations) to discriminate against an employee or job applicant on the basis of certain 'protected characteristics'[4]. These are:

- Age
- Disability
- Gender reassignment
- Marriage and civil partnership
- Pregnancy and maternity
- Race
- Religion or belief
- Sex
- Sexual orientation.

In Australia, workers are protected by federal and state anti-discrimination laws. At the federal level, these include the Australian Human Rights Commission Act (1986), Age Discrimination Act (2004), Disability Discrimination Act (1992), Racial Discrimination Act (1975) and Sex Discrimination Act (1984). Under federal anti-discrimination law an employer may be legally responsible for discrimination and harassment which occurs in the workplace or in connection with a person's employment unless it can be shown that 'all reasonable steps' have been taken to reduce this liability.[5] The Equal Opportunity for Women in the Workplace Act (1999) also requires all private sector companies, not-for-profit organisations, non-government schools, unions, group training companies and higher education institutions with 100 or more people to establish a workplace program to remove the barriers to women entering and advancing in their organisation.[6]

Other countries also have various legislation that prevents organisations from discriminating against individuals in selection for employment on grounds of race, sex, marital status and disability (Loosemore *et al.*, 2003). See **Table 7.1** for several examples of international equality legislation.

The construction sector has also voluntarily implemented its own equality and diversity agenda, largely in response to the business case for equality. In 1996 an industry-backed taskforce was established to explore equality issues in the UK construction industry (CIB, 1996); in 2000 a diversity toolkit was launched in the UK to help organisations improve their equal opportunities practices (Constructing Excellence, 2004). In Australia, there have been few industry-wide commitments to diversity and equality. An exception to this is with regard to Indigenous Australians. The NSW Government, for example, has introduced guidelines on Aboriginal participation in construction (NSW Government, 2007) to support and promote employment opportunities for Aboriginal people on government construction projects.

Table 7.1 International examples of equality legislation

Country	Key legislation
United Kingdom	• Equal Pay Act 1970 • Sex Discrimination Act 1975 • Race Relations Act 1976 • Disability Discrimination Act 1995 • Equality Act 2010
Australia	• Racial Discrimination Act 1975 • Sex Discrimination Act 1984 • Australian Human Rights Commission Act 1986 • Disability Discrimination Act 1992 • Age Discrimination Act 2004
South Africa	• Constitution of the Republic of South Africa 1996: Bill of Rights • Employment Equity Act 1998
Europe	• Equal treatment in employment and occupation (Directive 2000/78/EC) • Charter of Fundamental Rights 2007 • Framework decision on combating racism and xenophobia (Council Framework Decision 2008/913/JHA)
Denmark	• Prohibition of Discrimination in the Labour Market Act 2004.

Source: www.non-discrimination.net

However, legislative requirements, while valuable in protecting people from discrimination, have to-date, been inadequate in eliminating inequality (Bagilhole, 2009) and are not effective in changing attitudes (Loosemore *et al.*, 2003). Dainty *et al.* (2007) propose that government policy has done little to regulate industry employment more generally, either through direct labour market governance or via public sector procurement influence. Of particular concern has been the way contractors have systematically sought to avoid statutory responsibilities through bogus self-employment. Caplan (2007) argues that although most medium and large organisations now have equal opportunities policies, they are rarely seen as impacting on their core business. This is also problematic given that the majority of construction firms are small businesses (the fragmentation of the construction sector is discussed further below). Equally important is the distinction between Equal Opportunities and Diversity Policies. Kirton and Greene (2009) have argued that, in the UK the discourse of diversity management has replaced that of equal opportunities. The latter has more of a focus on the business case of diversity. However, the case studies presented here suggest that construction employers may not differentiate between the two. However, both do utilise the liberal approach whereby the business case is emphasised (Jewson and Mason, 1986).

Equality, diversity and inclusion

The majority of the construction sector is white, male and able-bodied and even in areas where this is not the case, minority workers are often treated poorly, as will be demonstrated below. While statistical data is generally poor, evidence in the UK and Australia indicates that women, ethnic minorities, people with a disability and older people represent less than one in seven construction workers (**Table 7.2, Table 7.3, Table 7.4**). Comparative data is also scarce, largely as a result of international differences in defining minority groups. Research also shows that the under-representation of minority groups and division of labour in construction exists elsewhere, including for example: women in Nigeria (Adeyemi *et al.*, 2006), Denmark, Italy, the Netherlands and Spain (Byrne *et al.*, 2005), Zanzibar (Eliufoo, 2007), Bangladesh and Thailand (Hossain & Kusakabe, 2005), Singapore (Ling & Poh, 2004); migrant workers in Denmark, Italy, the Netherlands and Spain (Byrne, *et al.*, 2005), Malaysia (Abdul-Aziz, 2001); ethnic minorities in Denmark, Italy, the Netherlands and Spain (Byrne *et al*, 2005); and, disabled workers in the Netherlands (Clarke *et al.*, 2009).

Table 7.2 Diversity in the Australian construction sector

	All employed	Employed in construction industry
Total (n.)	9,104,187	709,843
	%	%
Women	46.1	13.5
Aged 55 years or older	15.1	13.6
Indigenous	1.3	1.2
profound or severe disability	0.6	0.3
Migrants (people born overseas)	23.5	20.7
Migrants who speak a language other than English at home	10.1	7.4
Migrants who are not proficient in English	0.9	0.9

Note: People with a profound or severe disability are defined as needing help or assistance in one or more of the following activities because of a disability, long term health condition or old age: self-care, mobility, communication. Migrants are those people born overseas and intending to stay in Australia for at least one year (may have migrated in any year). Persons who are not proficient in English are those that stated they spoke English 'not well' or 'not at all').

Compiled using data from ABS (2006) Census of population and housing.

Table 7.3 Diversity in the construction sector in England and Wales

	All employed	Employed in construction industry
Total (n.)	22,441,498	1,515,996
	%	%
Women[1]	45.9	10.1
Aged 55 years or older[1]	13.4	15.4
Ethnic minorities (non-white)[1]	6.9	2.8
Long-term disabled[2]	12.8	13.1

Note: Long-term disabled includes those who have a long-term health problem or disability which has a substantial and long-term adverse effect on their ability to carry out normal day-to-day activities and/or people with a long-term health problem or disability which affects either the kind of paid work they might do, or the amount of paid work they might do.

Compiled using data from: [1] *ONS (2001) Census;* [2] *Meager and Hill, 2005, derived from LFS, 2005.*

Table 7.4 Diversity in the UK construction sector

	Employed in construction industry
Total (n.)	2,278,373
	%
Women	10.0
aged 55 years or older	17
ethnic minorities	4.1

Compiled using data from Labour Force Survey (2009) and Construction Skills (2008).

This section provides an overview of minority groups currently employed in construction, including women, ethnic minorities, migrant workers, people with a disability, and older people. However, it is important to recognise the heterogeneity within and between these groups. For example, workers may belong to multiple minority groups (e.g. ethnic minority woman), and as a result may face more or less disadvantage. Furthermore, the experiences of minorities working in construction will themselves be diverse and are likely to be associated with their role in the construction sector (e.g. the job they do), the other people they work with (e.g. the diversity of other team members), and the organisation they work for (e.g. organisational culture).

Women

Women remain numerically and hierarchically under-represented in the construction workforce (Powell *et al.*, 2010). Women represent 10 per cent of the UK construction workforce (**Table 7.4**) and 13.5 per cent of the Australian construction workforce (**Table 7.2**). They also represent 18 per cent of civil engineering students and 31 per cent of architecture, building and planning students in the UK (HESA, 2009). This is clearly below average with women representing 46 per cent of all those in employment and 57 per cent of all university students. Furthermore, while the industry has sought to address the under-representation of women, through various initiatives, progress to date has been relatively slow and inconsistent.

Research into the persistence of gender inequality in construction has investigated barriers and solutions to women's recruitment, retention and progression in the industry. Much of this has focused on the cultural and structural barriers women face in construction education and occupations (e.g., Dainty & Bagilhole, 2006; Fielden *et al.*, 2001; Greed, 2000; Lingard & Francis, 2006; Watts, 2007; Whittock, 2002).

Retaining women who are already engaged in the construction sector is also problematic. Bennett *et al.* (1999), for example, found that only 68 per cent of female professionals were committed to remaining in the industry compared to 87 per cent of male professionals. There are therefore real concerns that women's underachievement in construction may lead them to leave the sector (Loosemore *et al.*, 2003).

Watts (2007) suggests the low proportion of women in the sector is a result of the continued dominance of masculine stereotypes based on a male career model and a culture of long hours and presenteeism. Watts also maintains that women employed in the sector face various difficulties and barriers which result in their segregation within the industry. Such segregation is manifested both vertically, with women over-represented in more junior roles, and horizontally, with women over-represented in the public sector (Fielden *et al.*, 2001). Dainty *et al.* (2000) also found that women are likely to progress through their careers at a slower rate and face more obstacles than their male colleagues. Thus, while the industry has sought to address the under-representation of women, many within the sector seem reluctant to change (Fielden *et al.*, 2001), and it remains 'culturally, normatively and numerically male-dominated' (Watts, 2009:39).

One area which remains under-explored is gender, rather than sex. Further work is required to understand how experiences within the sector may be different for subordinated masculinities and femininities. For example, it is naïve to treat women and men as homogeneous groups; while some women are likely to relish working in a male dominated culture such as construction, it is equally likely that some men in the industry find the culture problematic. With some exceptions, much of the extant literature in the field is empirical rather than utilising sociological theory to understand the persistent marginalisation of women and certain groups of men. Theories such as hegemonic masculinity (as developed by Connell, 1982) and gender performativity (Butler, 1990, 2004) could be applied within the sector to understand how gender norms manifest within the sector and to help research move away from an essentialist view of gender.

Ethnic minorities

Census data in the UK indicates that less than three per cent of the construction workforce is from an ethnic minority background, much lower than average (**Table 7.3**; **Table 7.4**). A number of qualitative studies have also explored the experiences of ethnic minority workers in construction (see for example, Ahmed *et al.*, 2008; Caplan & Gilham, 2005; CEMS, 2005). In Australia, Indigenous Australians are employed in the construction sector at a similar to rate to other industries (**Table 7.2**). Beyond this, however, there is a lack of quantitative and qualitative data on the experiences of Aboriginal and Torres Strait Islanders in the Australian construction industry.

Caplan *et al.* (2009) found that in the UK ethnic minority students are reasonably well-represented on construction-related courses in further and higher education, yet remain under-represented in industry. Evidence suggests that ethnic minority students are less likely than their white counterparts to move from construction higher education to industry (CEMS, 2005). This is largely because ethnic minority students are less likely than others to complete their studies and more likely to seek employment outside the industry (CABE, 2004; CEMS, 2005). This may be a result of ethnic minority students receiving poor support within the university environment. Caplan and Gilham (2005), for example, found ethnic students to be fearful of seeking support and that when support was provided it often created 'us and them' divisions, although evidence of overt discrimination appeared rare (see also Caplan, *et al.*, 2009). Caplan and Gilham (2005) also found that ethnic minorities felt they were denied opportunities and were subject to negative assumptions such as, poor English, lack of communication skills, lack of practical experience, an unwillingness to travel and hypersensitivity. There was also a lack of trust and appreciation between academics and students that existed as an invisible barrier. Ethnic minority students also report a lack of access to the kind of social networks which are perceived as necessary for entry to the profession (CABE, 2004). They may lack the social capital necessary to socialise into the profession (Sang *et al.*, 2010b).

Even among companies that had developed equal opportunities policies and achieved better representation, minority groups rarely achieved this at 'more senior, desirable and powerful positions' (Caplan & Gilham, 2005:1010). It can also be difficult for ethnic minorities to 'fit in' and conform to construction cultures and norms; this can be particularly difficult if resentment or racism is unchallenged or subtle. They also found that ethnic minorities that have been successful face huge pressures, with potential to positively influence the white majority, but at risk of reinforcing racial stereotypes.

Loosemore and Chau (2002), in their study of Asian operatives in the Australian construction sector, found that 40 per cent of interview respondents felt they had suffered some discrimination and their employment had been detrimentally affected by their race. This was despite most companies having equal opportunities policies in place. Operatives reported coping with racism by ignoring it or using social support mechanisms and stated that racism seemed to create a socially fragmented workplace. They also suggested that racism negatively impacted on productivity by reducing levels of co-operation, goodwill and open communications, as well as adversely affecting morale and creating resentment in the workplace. Those facing racism were reluctant to report it to managers. Such reluctance to report incidents makes it difficult for employers and human resources managers to tackle discrimination and harassment. More recently, Loosemore *et al.* (2010) found that racist graffiti and humour were widespread on Australian construction sites with over half of those they surveyed acknowledging this.

Migrant workers

The construction industry has long relied on migrant labour which is relatively cheap (TUC, 2006). In comparison to other countries, the Australian construction sector has absorbed a relatively high proportion of migrants into its workforce (Loosemore & Chau, 2002). Migrant workers have often been used by the sector to fill skills shortages (Schellekens & Smith, 2004). The HSE (2010) estimates that approximately six per cent of the UK construction workforce is from overseas, while Lillie and Greer (2007) and Chan *et al.* (2010) suggest that the share of migrants working in the UK construction industry is as high as ten per cent. In Australia, it is estimated that around one in five workers in the construction industry are born overseas, with half of these from non-English speaking countries (DIAC, 2009; Loosemore *et al.*, 2010; **Table 7.2**). Determining the precise numbers of migrants workers employed within the construction industry is problematic. Dainty *et al.* (2007) point out that self-employed migrant workers have no need to apply for a work permit and so are not included in official statistics. Furthermore those employed unofficially (undeclared) are also not included in official statistics. This problem is exacerbated by the often transitory status of many migrant workers (van den Brink & Anagboso, 2010).

In some instances, migrant workers may contribute to the diversity of the workforce, particularly if they are used to fill skills shortages (e.g., Clarke & Gribling, 2008). However, this is not to say that migrant workers are treated equitably or inclusively and they are often segregated in particular areas of the industry.

A study by the TUC (2006) into the experiences of migrant construction workers in the North East of England highlighted several areas of concern. These included migrant workers living in poor quality housing which was often owned by their employer; an inability to secure bank accounts (essential for integration into the UK workforce, ConstructionSkills, 2010) due to employer reluctance to supply necessary documentation; the absence of pay slips or contracts of employment; low wages, often below the minimum wage; employer hostility including violence; and language barriers. Language barriers are a particular concern for health and safety. For example, when migrants have English as a second language it is unclear whether they understand signage on site and verbal instructions (Schellekens & Smith, 2004; Loosemore *et al.*, 2010). Local workers were generally receptive to the problems faced by migrant workers, but the report expressed concerns that if the number of migrant workers continues to increase, this support may not continue (TUC, 2006). There appears to be little contact between second language speakers (migrants) and trade unions with many having little understanding of the role of unions and union representatives reporting little contact with migrants (Schellekens & Smith, 2004). This has potentially important consequences for the treatment of migrant workers. The report by Schellekens and Smith (2004) focuses on the experiences of second language speakers who are legally employed within the UK construction industry. The treatment of those engaged within the black market economy remains unexamined.

People with a disability

In the United Kingdom people with a disability are protected by the Disability Discrimination Act (1995 as amended in 2005). Under this Act disability is defined as 'a mental or physical impairment which has a substantial and long-term adverse effect on a person's ability to carry out normal day-to-day activities'. The Act makes it unlawful to discriminate against a person with a disability in terms of employment as well the provision of goods and facilities. Since October 2010, those with a disability are also protected by the new Equality Act (see above). There is very little research exploring the experiences of those employed within the sector that have a disability (Gale & Davidson, 2006).

While little is known about the situation of those with disabilities, there are particular implications for human resource managers. In the UK those who employ 15 people or more must make what is termed 'reasonable adjustments' in or to the working environment. This may include reallocation of tasks or considering flexible working hours. All employers are required to ensure that they do not treat an employee or applicant in a less favourable way because of their disability (Foot & Hook, 2002).

People with a long-term disability are more likely to be unemployed (4.1%) or economically inactive (46.3%) than people with no disability (3.6% and 16.1% respectively; Meager & Hill, 2005). Disabled workers also face multiple disadvantages when they are able to participate in the labour market (Newton & Ormerod, 2005). Within the construction sector, comparative data on the number of people employed with a disability or long-term health condition is lacking. This is due to a lack of formal monitoring of disability within companies and problems of definition. For example, in England and Wales, approximately 13.1 per cent of the construction workforce has a long-term disability (slightly above average), while in Australia 0.3 per cent of the construction workforce has a profound or severe disability (below average) (see **Table 7.2** and **Table 7.3** for further information).

The construction sector can itself be seen as a major disabler of individuals, with construction workers at greater risk of developing certain health conditions than average (Brenner & Ahern, 2000). Common health conditions affecting construction workers include back injuries, injuries due to hits or falls, respiratory infections, arthritis and hearing deficiencies, (*ibid.*; see also Clarke *et al.*, 2009). Lingard and Saunders (2004) found that in Victoria, Australia, few construction companies had rehabilitation or return-to-work programmes for injured workers. They also found that small construction firms particularly lacked the resources and knowledge to support injured and disabled workers.

Another barrier facing disabled workers is employer attitudes. Many people with disabilities can work safely on construction sites with suitable support and physical barriers to working in office-based roles should be minimal (Lingard & Saunders, 2004). However, Lingard and Saunders (*ibid.*) for example, found that employers were hostile and suspicious of workers motives in occupational rehabilitation, which may result in workers feeling blamed, discouraged or punished after an injury. Newton and Ormerod (2005:1077) found that while employers said they would recruit disabled workers, the employee would need 'to be "able to maintain frequent unaided mobility bother around the UK without the use of public transport and around a wide variety of construction site environments, and to be able to sustain a key small team role between 7.30 a.m. and 6.00 p.m. including breaks" (contractor survey interviewee response)'. This is clearly problematic for some people with a disability, as well as many non-disabled workers. Newton and Ormerod (*ibid.)* also found that employers tend to adopt a narrow definition of disability, limited to physical and sensory impairments; do not monitor the impact of their policies on disabled workers; are more likely to employ people with a disability in an office role rather than a site-based role; and, adopt a piecemeal approach to disability, negotiating adjustments on an individual need basis.

Older workers

The most recent data available suggests that in England and Wales older workers aged 55 years or over are slightly over-represented in the construction workforce (15.4% compared to an average 13.4%; **Table 7.3**). On the other hand, in Australia older workers are slightly under-represented (13.6% compared to an average 15.1%; **Table 7.2**). Literature exploring discrimination against construction employees on the basis of age appears to be sparse. There is some anecdotal evidence that young construction professionals, or those who appear young, may experience negative attitudes from site workers (Sang, 2007). Data from other sectors suggests that stereotypes about older workers can significantly and negatively affect employees' willingness to work with older workers (Chui *et al.*, 2001). However, anti age discrimination policies can help to overcome these problems.

Summary

In summary, the construction industry's approach to equality and diversity has largely centred round a business-case argument rather than an ethical one based on justice and inclusion. This has resulted in minimal change to the workforce, over the last few decades, which remains predominantly white, male and able-bodied. In some areas, including some countries, regions and sub-sectors of the industry, minorities are represented in higher numbers. However, these minorities are often segregated, vertically and horizontally, and work in 'silos', reflecting a lack of inclusion. This approach to equality and diversity paints an unappealing picture of the industry, as exclusive and discriminatory, and is likely to make recruitment and retention of minority workers challenging. Some of the barriers to equality, diversity and inclusion are discussed further below.

Case Study 7.1

May Gurney's equality and diversity policy

May Gurney is an infrastructure support services company and its group office is based in Norwich (Norfolk, UK). It employs 4,700 people across a range of disciplines including construction. The company's vision is to 'Be the Best', which focuses on three main elements:

- *'Best customer service:* increasing focus on building long-term relationships with our customers, understanding their needs and becoming ever more important to their business. This includes working hard for customers' customers – members of the public.
- *Best performance:* delivering promises to customers by continually improving the way we do things and meeting stakeholders' expectations.
- *Best place to work:* recruiting, keeping and developing talent is crucial to success. Central to this is creating a working environment where people feel happy and motivated, supported and developed.'

As part of this vision May Gurney 'aims to create a working culture that respects the value of differences among colleagues ... to ensure equality of access, process and outcomes within an environment that is inclusive, open and fair'.

The information for this case study is taken from the May Gurney website and an interview with Liz Bond (Starting out Officer and member of Equality and Diversity working group) and Ellie Neal (Diversity Group Lead).

May Gurney has a single Equality and Diversity policy which is publically available from their website. This policy has been adapted to fit the requirements of the 2010 Equality Act.

Examples of the policy practice

- Disability is accommodated on a case-by-case basis, e.g., employing sign language interpreters for staff who have hearing impairments.
- Work is ongoing to ensure language is not a barrier to full participation, for example, translating staff surveys into Polish.
- Coaching programmes are available for all staff to learn leadership skills.
- Open positions are advertised in a range of media.
- New equality monitoring form to measure effectiveness of policies and procedures.
- Compulsory equality and diversity training being delivered to managers. Training aims to help managers appreciate the differing views, e.g., of women. This training will then be rolled out to all staff (including those on site) via 'Ambassadors'.
- Flexible working policies in place open to all and infrastructure to support flexible working (mobile 'phones, laptops).

Benefits of the policy to May Gurney

- Staff morale and productivity – the aim is to ensure that all staff can fulfil their full potential.
- Commercial – e.g., securing bids for public sector contracts (which include social clauses) and reputation of the company.
- Legal requirement – to ensure compliance with legislation.

For further information visit May Gurney's website at: www.maygurney.co.uk/pages/careers_equal_op.html

Work–life balance

This section of the chapter focuses on WLB issues, which have been shown to be a concern for many employees within the sector, but can often have a major impact on organisational culture and subsequently minority groups. We consider the effectiveness of the construction sector in supporting employees to achieve WLB, since previous research indicates that few construction companies implement flexible work practices, despite the documented negative impact of poor WLB on employees' mental and physical health. Although certain demographic groups may be particularly vulnerable to poor WLB, this issue is important to all employees and should therefore concern all employers within the sector if they wish to recruit and retain a highly skilled workforce. Adopting WLB strategies may also stimulate more general cultural change (Lewis & Lewis, 1996), which is likely to positively impact on the experiences of minority groups.

Measures to support WLB may include leave arrangements (e.g. maternity, paternity, parental, bereavement, compassionate or unpaid leave); flexible working arrangements (e.g. part-time hours, flexible start and finish times, compressed working week, reduced hours, home working) and workplace facilities (e.g. childcare, counselling, stress management; Callan, 2007). Organisations that fail to address WLB usually assume a masculine work model, based on an 'ideal worker', unencumbered by domestic responsibilities (Gornick & Meyers, 2009; Williams, 2001). However, this model is increasingly out-dated for most workers, but particularly for those with caring responsibilities, who are primarily women.

The project-based culture in construction is argued to result in a working environment which encourages long working hours and presenteeism and hence poor WLB. Poor WLB has been linked to a range of negative outcomes for individuals, namely, poor health and wellbeing and organisations with increased voluntary turnover. Working on a project-by-project basis may make it difficult for employees to take time off. The traditional competitive tender approach to construction procurement has led to pressures on those working on projects. Those pressures include a need to complete the project within a given time and budget. As a consequence, employees are often required to work long hours to meet high workloads and strict budgetary constraints. This situation can be worsened if employees need to travel long distances to reach a site, thus lengthening the working day (Loosemore *et al.*, 2003). Many researchers have argued that time pressures and long working hours are damaging to those working within the sector (Sutherland & Davidson, 1993; Lingard & Sublet, 2002; Lingard, 2003; Haynes & Love, 2004; Love & Edwards, 2005).

The need to meet the tight time and financial constraints resulting from competitive tendering means that construction employees can face a potentially damaging high workload (Sutherland & Davidson, 1993; Lingard & Sublet, 2002; Lingard, 2003; Haynes & Love, 2004). Like long working hours, workload is associated with relationship conflict (Lingard & Sublet, 2002), poor mental health (Sutherland & Davidson, 1993) and other measures of wellbeing (Haynes & Love, 2004). Lingard (2003) identified a relationship between high workload and burnout. Burnout can have serious consequences for an organisation, as it has been associated with high staff turnover (Lingard, 2003), a matter which needs to be taken seriously in an industry which is experiencing a skills shortage and possibly a recruitment crisis (Dainty & Edwards, 2003) despite the current recession (Construction Industry Council, 2010).

On the next page we present a Case Study of an architectural practice in the UK which has successfully implemented a range of WLB policies. This work has been rewarded, not only with low staff turnover and low sickness absence rates, but also with national awards. In October 2010, LSI Architects was placed third nationally in the list of Good Employers in the Construction Industry.[7]

Case Study 7.2

Flexible working and equality and diversity at LSI Architects

LSI Architects is an architectural and interior design practice with offices in Norwich (Norfolk, UK) and London (UK). The practice employs around 50 staff. The practice ethos is that the architect is '*well-placed to provide a lead in the procurement process, and a conviction that high quality service and design usually arise from collaborative effort*'. In addition they target sustainable projects. LSI architects has won a number of prizes for its employment practices. In 2010 it was named one of the *Building Magazine*'s Top 5 construction employers in the UK. Through ongoing consultation between the partners and employees (staff surveys) a range of flexible working benefits have been put into place. These benefits are regularly reviewed to ensure they meet the needs of both the practice and the employees. The following information is taken from the LSI Architects website and an interview with Ben Goode (partner, Finance).

Policies and practices

- Staff are limited to a 37.5-hour week, although how these hours can be worked is flexible. Some staff choose to start work at 7 a.m. or 10 a.m. in order to miss heavy city centre traffic. This is then matched by leaving work earlier or later. Exact working hours must be coordinated with project needs and negotiated with the rest of the project team. To date, no requests for flexible working have been turned down.
- Staff complete daily electronic time sheets which are submitted weekly and monitored at a finance meeting. If staff exceed their working hours then steps are taken to bring in additional help on a project.
- Infrastructure is in place for employees in order to work from home (e.g. smart 'phones and laptops with access to practice portal).
- A staff handbook provided and reviewed regularly to ensure policies adhere to legislative changes e.g. the Equality Act 2010.
- Robust Equality and Diversity policies in place (recently modified to ensure compliance with the Equality Act 2010).
- Reasonable adjustments made for those with a disability. This includes flexible working requests and regular workstation assessments.

Benefits to practice and employees

- Limiting working hours to 37.5 per week reflects steps taken to prevent burnout.
- Social policies such as flexible working are beneficial when bidding for public sector contracts.
- Very low staff sickness and low staff turnover.
- As a practice based in Norwich where there are no nearby architectural schools recruitment can be problematic. Ensuring that they are able to offer a range of benefits, beyond salary, enables LSI Architects to recruit the highest calibre employees who share the practice's ethos.

For further information visit LSI Architects' website at: www.lsiarchitects.co.uk

Barriers to equality, diversity, inclusion and work–life balance

The barriers to EDI and WLB in the construction sector identified by previous research generally focus on the structure and cultures of the industry. Cultural barriers include the persistence of stereotypes about minority groups, direct and indirect discrimination and the assimilation of employees into the culture, while structural barriers include the continued importance of networks in the industry, recruitment and training practices, long and inflexible working hours, dependence on formal policies to instigate change and the fragmented nature of the industry. Each of these factors is discussed further below, although they are not mutually exclusive.

Persistent stereotypes and discrimination

Powell *et al.* (2010) suggest that resistance to cultural change in construction is promoted by the entrenched nature of gendered stereotypes, which mean that women's retention and progression in the industry remain problematic despite initiatives to improve their situation. Agapiou (2002), for example, found that male construction workers thought that women involved in the structural aspects of building work would compromise their femininity, as the 'hard and dirty' image of the building site did not match their perceptions of suitable work for women. Loosemore *et al.* (2010) found that tensions existed between ethnic groups on Australian construction sites as a result of stereotypes which characterised Lebanese-Australians as aggressive and Asian-Australians as having disregard for safety standards. Such deep-seated stereotypes and attitudes, coupled with the historical and cultural dominance of white, able-bodied men can result in direct, indirect and institutionalised discrimination within the construction sector (Craw *et al.*, 2007).

Networks

Those from minority groups working in construction may also find themselves excluded from informal networks which are often critical in terms of gaining access to the workplace (see also recruitment below), procurement and sub-contracting, relevant work experience and career development (Loosemore *et al*, 2003; Craw *et al*., 2007). CABE (2004), for example, found that ethnic minority workers may lack the social capital necessary to socialise into the construction profession. As informal networking often takes place outside the workplace, when some workers are less likely to be available or feel uncomfortable attending (for example, social events such as drinking after work or playing sport), opportunities to develop networks may be limited. Some attempts have been made to establish networking opportunities for minority groups in construction, for example women's networks. However, these networks usually comprise only the minority group, which while providing support, do not necessarily challenge the existing power structures or provide the group with access to those in decision-making positions.

Recruitment and training practices

Recruitment in construction often relies on informal and personal contacts (Druker & White, 1996), particularly for skilled labour on construction sites (e.g. Ness, 2010), which can work as a powerful form of exclusion. Clarke and Gribling (2008) suggest that the 'exclusivity' of the construction sector is shaped by the provision of inappropriate training, and employment and working conditions, notably traditional practices such as old-style apprenticeships, craft-based skill structures, an itinerant workforce, and intensive deployment of labour. Furthermore, inadequate links between training provision and employers' needs, limited apprenticeships and work experience placement result in barriers for groups in formal training to access work experience and employment opportunities (Craw *et al*., 2007).

EDI and WLB policies

While formal EDI and WLB policies are valuable, they do not automatically lead to change if they are not implemented appropriately (e.g., Barnard *et al*., 2010; Clarke & Gribling, 2008). Barnard *et al*. (2010) argue that cultural norms can make it difficult for men and women to take-up formal opportunities for family-friendly working. While equal opportunities policies and diversity management strategies are now the norm in many organisations, they can have a limited effect on masculine workplace cultures which value total availability and commitment from employees.

Added to this, Caplan *et al*. (2009) suggest that while there is general acceptance for diversity among industry leaders, the implementation of equality and diversity policies is on the whole restricted to large companies, rather than the SMEs that proliferate in construction. Furthermore, they argue, construction firms misunderstand the importance of equality targets and are therefore hesitant to implement them.

Procurement practices

Public authorities can use procurement to implement and monitor equality and diversity practices. However, evidence suggests that authorities are making minimal progress in terms of using procurement to promote equality, largely due to lack of coordination and consistency (Craw *et al.*, 2007). For example Duncan and Mortimer (2007:1291) found that 'few organisations made efforts to ensure that the contractors and consultants they engaged actually implemented equality and diversity within their organisations.' Caplan *et al.* (2009) also suggest that traditional procurement methods and the use of approved lists may contribute to indirect discrimination against ethnic minority contractors.

Long work hours

Dainty and Lingard (2006) have found that the requirement to conform to the notion of an 'ideal worker' also militates against equality. Within construction, long, irregular shifts, work hours and travelling times result in a preference for mobile workers (Craw *et al.*, 2007). This can impose restrictions on women's careers in construction in comparison to men's, as women are often forced to make a choice between work and family (as well as others with responsibilities outside the workplace). Evidence also suggests that employees who do not demonstrate total commitment and availability (by working long hours) or who take advantage of WLB policies, are likely to face subtle discrimination, suspicion and marginalisation within their organisation (Bagilhole, *et al.*, 2007; Dainty & Lingard, 2006; De Graft-Johnson, Manley, & Greed, 2005; Fielden *et al.*, 2000; Watts, 2009). This also highlights the rhetorical nature of many formal policies directed at improving equality.

Fragmentation

Craw *et al.* (2007) found that the fragmented nature of construction has impacted on diversity, by diffusing the diversity agenda and enabling individuals and organisations to pass responsibility to others. The competitive tendering processes which emphasise costs and leave few resources for diversity measures; and the prevalence of self-employment and temporary agency workers can be seen to limit responsibility for workforce development. **Table 7.5** shows that the majority of professional service firms in UK construction employ fewer than six people meaning that most will

Table 7.5 Professional service firms in UK construction industry operating as sole traders and micro businesses

	Sole trader	Fewer than six people	Total (fewer than six people)
	%	%	%
Architecture	35	42	77
Civil engineering	20	36	56
Building services	20	41	61
Quantity surveyors	28	37	65
Surveyors	29	47	76
Managers	22	44	66

Compiled using data from Labour Force Survey 2009.

have no formal human resource department and that those with responsibility for undertaking professional work also have HR responsibilities.

As the industry is characterised by small companies, high self-employment and short-term projects, business demands and the need to maintain a competitive edge are often valued at the expense of WLB for workers.

Assimilation into the industry

Bennett *et al.* (1999) claim that women who seek a career in the construction industry are socialised into its culture through the education system and appear actively to seek that culture. Gale (1994) describes gender values as a continuum ranging from male to female and suggests that women holding similar values are attracted to similar occupations. Bennett *et al.* (1999) do, however, concede that the reverse is also true: many women reject the construction culture as unacceptable, as do many men.

Powell *et al.* (2010) found that women in construction subscribed to many of the gender stereotypical views of other women and who could be successful in the construction sector. These attitudes have also been described in existing research as women's assimilation or professionalisation into the masculine cultures they experience daily (see for example, Dryburgh, 1999; Faulkner, 2006). While there is much debate around whether this is a deliberate strategy women adopt to cope in masculine environments or simply a result of 'becoming' a construction professional, it does not challenge the status quo and makes cultural change unlikely. Any career success among such women is unlikely to promote the interests of women in the sector generally (Greed, 2000). It also raises questions about the concept of a 'critical mass': the idea that once there is a sufficient proportion of women in engineering, traditionally masculine cultures will no longer prevail. Williams and Emerson (2001), for example, suggest that if women were to account for at least 30 per cent of the workforce the existing cultures may be challenged (see Powell *et al.*, 2006 for further discussion of critical mass).

Recommendations for change

The construction industry faces a number of structural and cultural barriers. These barriers appear to be particularly steadfast and it is likely that they need to be challenged in more radical and/or innovative ways than has been the case previously, before there can be significant progress towards EDI for minority groups and effective WLB. Simply increasing the numbers of minority groups without addressing these barriers is unlikely to bring about any long-term improvements (Loosemore *et al.*, 2003).

Having identified the key barriers to progress in terms of EDI and WLB in construction, and using examples from the extant construction and HRM literature, we now describe some recommendations for change for construction organisations, public sector authorities and the construction sector more generally. For each of these recommendations, the cost implications for employers should be considered.

In order to understand the current profile of their workforce, employers should consider monitoring equality by collecting data on the diversity of employees and subcontractors. Doing so will enable employers to establish if hierarchical and vertical segregation is present within their workforce. Understanding the current profile of the workforce would also enable diversity targets to be established. In addition, construction employers should remember their responsibility across the supply chain. For example, it may be appropriate to insist that sub-contractors monitor diversity or they may implement diversity awareness raising and training across management and the workforce, including subcontractors (Craw *et al.*, 2007).

To determine the effect of organisational policies on different sectors of the workforce, private sector employers should consider undertaking equality impact assessments. Such assessments can allow action to be taken where necessary to promote diversity or address inequality (Bond *et al.*, 2009). Public sector employers in the UK are legally bound to undertake equality impact assessments on policies, for example, to determine how a compassionate leave policy may have differing effects on men and women. While private sector organisations are not legally obliged to do so, they may find these assessments useful to understand how their policies and practices affect the various demographic groups they employ. Many guides now exist to aid companies in doing this (e.g., Acas, 2009).

Employers may also wish to appoint a diversity champion who can take forward and promote relevant policies within the organisation (e.g., Bond *et al.*, 2009). As shown in the case study, this approach is being utilised by May Gurney with the use of equality and diversity Ambassadors. It is important that employers are aware that there may be a cost (both financial and to the diversity manager) of undertaking such a role. Data from other sectors suggests that Diversity Champions often undertake this kind of work in addition to their regular workload (Kirton and Greene, 2009). As such they are more embedded within organisations than equality officers or diversity managers often are. However, undertaking such a role can lead to marginalisation within the organisation and a sense of personal failure if diversity measures are not successful (*ibid.*).

In addition to nominating diversity champions, there is a need for transparent and inclusive methods of recruitment, including advertising positions as widely as possible and ensuring that interview panels are diverse (Caplan & Gilham, 2005). Another approach may be for construction companies to utilise the programmes of external diversity organisations. For example, Stonewall, the UK-based gay and lesbian rights organisation has a list of Stonewall Diversity Champions, which are organisations that have met a set of criteria for promoting equitable working environments for lesbian, gay and bisexual staff (Stonewall, 2010). The Stonewall website has a jobs page where Stonewall Diversity Champions can advertise vacancies. This list includes construction employers (in 2010).

At present there is evidence that utilising work–life balance policies may be detrimental to an employee's career (or that perception exists). To counter this, employers should ensure that work–life balance (WLB) policies are seen as relevant to all workers, and that use of WLB strategies will not be detrimental to career advancement. WLB policies are also often thought of as women's issues. If WLB policies are to be successful, their use needs to be promoted among all employees, not just women and parents.

Conclusion

This chapter has shown that the construction sector has largely adopted a business case approach to equality and diversity, rather than an ethical one. Despite government regulation and the implementation of some industry self-regulation and diversity and work–life balance initiatives, the construction sector remains predominantly white, male and able-bodied. Where minorities such as women, ethnic minorities and migrant workers are better represented within the industry, they continue to be a minority and face a number of cultural and structural barriers that impede their development and opportunities for moving between sub-sectors or promotion. Many of these barriers have also hampered widespread acceptance of work–life balance policies and initiatives. These barriers include persistent stereotypes and discrimination, the predominance of informal networks and recruitment practices, a dichotomy between policy rhetoric and reality, procurement practices that fail to address equality, an expectation of total availability that results in long work hours and presenteeism as normative practice, the fragmented nature of the industry and the assimilation of minority groups to existing cultural norms. These steadfast barriers are challenging for individuals, organisations and the industry as a whole to overcome. Where best practice does exist, further research is required to identify the extent to which the practice has actually improved equality, diversity, inclusion and/or opportunities for greater work–life balance. It should also examine how such practices were implemented and how they might be applied in other contexts, such as in other organisations, different countries, or with different minority groups.

While we have made a number of specific recommendations for change for construction associations, public sector authorities and construction organisations and employers, we also argue strongly that it is time for dramatic cultural change. Such change needs both a top-down and bottom-up approach, supported by policy. Failure to do so may result in an unsustainable workforce. Further work is needed in particular areas. There has been considerable research exploring the experiences of women in the construction industry and their relative lack of career progression. A growing body of literature is beginning to address the experiences of ethnic minorities. However, it has become clear that the extant literature has not given voice to other marginalised groups within the sector, specifically, older (and younger) workers, lesbian, gay, bisexual and transgender employees and those with disabilities. In addition, class remains under-explored. Neither has it addressed how any of these characteristics might intersect. Further work is needed to understand how gender, ethnicity, disability/being able-bodied, sexual orientation, class and gender identity influence the experiences of those working within the sector. Doing so will help to identify how the sector can work to ensure equitable working environments for all its employees and how it can meet the needs of a diverse client base.

Notes

1 www.direct.gov.uk/en/Employment/Employees/Flexibleworking/DG_10029491
2 WISE (Women into Science and Engineering) is a UK-based initiative to encourage more girls to value and pursue careers in science, technology, engineering, mathematics and construction.
3 From 2011, there will be a new equality duty for public authorities in the UK, which will expand the existing duties to cover race, disability, gender, gender identity, religion/belief, age and sexual orientation.
4 See the UK Equality and Human Rights Commission (www.equalityhumanrights.com) for further information.
5 See the Australian Human Rights Commission (www.hreoc.gov.au) for further information.
6 See the Equal Opportunity for Women in the Workplace Agency, Australian Government (www.eowa.gov.au) for further information.
7 www.itmps.co.uk/digitaleditions/BLDGEGOCT10.html

References

Abdul-Aziz, A.-R. (2001) Foreign workers and labour segmentation in Malaysia's construction industry, *Construction Management and Economics*, 19(8), 789–98.

Abersoch, D. (2011) Women on Boards: Independent review, accessible at: www.bis.gov.uk/assets/biscore/business-law/docs/w/11-745-women-on-boards (accessed February 2012).

ABS (2006) *Census of Population and Housing*. Cat no. 2068.0. Canberra: Australian Bureau of Statistics.

Acas (2009) *Delivering equality and diversity*. London: Advisory, Conciliation and Arbitration Service.

Acas (2010) *The Equality Act: What's new for employers?* London: Advisory, Conciliation and Arbitration Service. Available at: www.acas.org.uk/CHttpHandler.ashx?id=2833&p=0 (accessed October 2010).

Adeyemi, A. Y., Ojo, O., Aina, O. O. and Olanipekun, E. A. (2006) Empirical evidence of women's under-representation in the construction industry in Nigeria, *Women in Management Review*, 21 (7), 567–77.

Agapiou, A. (2002) Perceptions of gender roles and attitudes toward work among male and female operatives in the Scottish construction industry, *Construction Management and Economics*, 20 (8) 697–705.

Ahmed, V., Pathmeswaran, R., Baldry, D., Worrall, L. and Abouen, S. (2008) An investigation into the barriers facing black and minority ethnics within the UK construction industry, *Journal of Construction in Developing Countries*, 13(2): 83–99.

Alasuutari, P. (1995) *Researching culture: qualitative method and cultural studies*, London: Sage.

Bagilhole, B. (1997) *Equal Opportunities and Social Policy: issues of gender, race and disability*, London: Longman.

Bagilhole, B. (2009) *Understanding Equal Opportunities and diversity: the social differentiations and intersections of inequality*, Bristol: Policy Press.

Barnard, S., Powell, A., Bagilhole, B. and Dainty, A. (2010) Researching UK women professionals in SET: A critical review of current approaches, *International Journal of Gender, Science and Technology*, 2(3): 361–81.

Bennett, J. F., Davidson, M. J. and Gale, A. W. (1999) Women in construction: a comparative investigation into the expectations and experiences of female and male construction undergraduates and employees, *Women in Management Review*, 14(7), 273–91.

Blanchflower, D. G. and Wainwright, J. (2005) An analysis of the impact of affirmative action programs on self-employment in the construction industry, Working paper 11793. Cambridge, MA: National Bureau of Economic Research. Available at: www.nber.org/papers/w11793 (accessed October 2010).

Bond, S., Hollywood, E. and Colgan, F. (2009) *Integration in the Workplace: emerging employment practice on age, sexual orientation and religion or belief*, Research report 36. Manchester: Equality and Human Rights Commission.

Bottero, W. (1992) The changing face of the professions? Gender and explanations of women's entry to pharmacy. *Work, Employment and Society*, 6(3), 329–46.

Brenner, H. and Ahern, W. (2000) Sickness absence and early retirement on health grounds in the construction industry in Ireland, *Occupational and Environmental Medicine*, 57: 615–20.

Brown, A. (1995) *Organizational culture*. Essex: Pearson Education.

Butler, J. (1990) *Gender Trouble*. London: Routledge.

Butler, J. (2004) *Undoing Gender*. London: Routledge.

Byrne, J., Clarke, L. and Van der Meer, M. (2005) Gender and ethnic minority exclusion from skilled occupations in construction: a Western European comparison, *Construction Management and Economics*, 23 (10), 1025–34.

CABE (2004) *Architecture and Race. A study of black and minority ethnic students in the profession*. London: Commission for Architecture and the Built Environment.

Callan, S. (2007) Implications of family-friendly policies for organizational culture: findings from two case studies, *Work, Employment and Society*, 21(4), 673–91.

Caplan, A. (2007) Access and inclusivity of minority ethnic people in the construction industry, A. Dainty, Green S. and Bagilhole B. (eds), *People and Culture in Construction: a Reader*. London: Routledge.

Caplan, A. and Gilham, J. (2005) Included against the odds: failure and success among minority ethnic built environment professionals in Britain, *Construction Management and Economics*, 23(10), 1007–15.

Caplan, A., Aujla, A., Prosser, S. and Jackson, J. (2009) *Race Discrimination in the Construction Industry: a thematic review*, Manchester: Equality and Human Rights Commission.

CEMS (2005) *Black and Minority Ethnic Representation in the Built Environment Professions*, London: Centre for Ethnic Minority Studies and Commission for Architecture and the Built Environment.

Chan, P., Clarke, L. and Dainty, A. (2010) The dynamics of migrant employment in construction: Can supply of skilled labour ever match demand? M. Ruhs and Anderson B. (eds), *Labour Shortages, Immigration and Public Policy*, Oxford: Oxford University Press.

Chandra, V. and Loosemoore, M. (2004) Women's self-perception: an inter-sector comparison of construction, legal and nursing professionals, *Construction Management and Economics*, 22 (9), 947–56.

Chui, W.C.K., Chan, A.W., Snape, E. And Redman, T. (2001) Age stereotypes and discriminatory attitudes towards older workers: an East-West comparison, *Human Relations*, 54 (5): 629–61.

CIB (1996) *Tomorrow's Team: women and men in construction*, London: Thomas Telford.

Clarke, L. and Gribling, M. (2008) Obstacles to diversity in construction: the example of Heathrow Terminal 5, *Construction Management and Economics*, 26 (10): 1055–65.

Clarke, L., Van der Meer, M., Bingham, C., Michielsens, E. and Miller, S. (2009) Enabling and disabling: disability in the British and Dutch construction sectors, *Construction Management and Economics*, 27 (6), 555–66.

Connell, R. W. (1982) Class, patriarchy, and Sartre's theory of practice, *Theory and Society* 11:305–20.

Constructing Excellence (2004) *Respect for People: a framework for action*, Report of the Respect for People Working Group, London: Constructing Excellence.

Construction Industry Council (2010) *The impact of the recession on construction industry professions. A view from the front line*, available at: www.cic.org.uk/newsevents/ImpactRecessionProfessionals-FrontLine.pdf.

ConstructionSkills (2010) *Focus on Migrant Workers*, available at: www.cskills.org/focus/migrantworkers.aspx (accessed October 2010).

ConstructionSkills (2010) *Construction Skills Network 2010–2014: Blueprint for UK Construction Skills 2010 to 2014*, Labour Market Intelligence. Available at: www.cskills.org/uploads/csn2010-2014national_tcm17-18127.pdf (accessed October 2010).

ConstructionSkills (2008) *Construction Industry Warnings on workforce time-bomb*. Available at: www.cskills.org/newsandevents/news/pastnews/workforce_time_bomb.aspx (accessed October 2010).

Coussey, M. and Jackson, H. (1991) *Making Equal Opportunities Work*, London: Pitman Publishing.

Craw, M., Clarke, L., Jefferys, S., Beutel, M., Roy, K. and Gribling, M. (2007) *The Construction Industry in London and Diversity Performance*, London: Greater London Authority.

Dainty, A. and Bagilhole, B. (2005) Guest Editorial *Construction Management and Economics*, 23 (10), 995–1000.

Dainty, A. R. J. and Bagilhole, B. (2006) Women's and men's careers in the UK construction industry: a comparative analysis, A. W. Gale and Davidson M. J. (eds), *Managing Diversity in the Construction Industry: initiatives and practices*, Taylor & Francis, London.

Dainty, A. and Edwards, D. J. (2003) The UK building education recruitment crisis: a call for action, *Construction Management and Economics*, 21(7), 767–75.

Dainty, A. and Lingard, H. (2006) Indirect discrimination in construction organizations and the impact on women's careers, *Journal of Management in Engineering*, 22(3), 108–18.

Dainty, A., Bagilhole, B. and Neale, R. (2000) A grounded theory of women's career under-achievement in large UK construction companies, *Construction Management and Economics*, 18(2), 239–50.

Dainty, A., Gibb, A., Bust, P. and Goodier, C. (2007) *Health, Safety and Welfare of Migrant Construction Workers in the South East of England*, London: Institute of Civil Engineers.

Dainty, A., Green, S. and Bagilhole, B. (eds) (2007) *People and Culture in Construction: A Reader*, London: Routledge.

De Graft-Johnson, A., Manley, S. and Greed, C. (2005) Diversity or the lack of it in the architectural profession, *Construction Management and Economics*, 23(10), 1035–43.

Devine, F. (1992) Gender segregation in the engineering and science professions: a case of continuity and change, *Work, Employment and Society*, 6(4), 557–75.

DIAC (2009) *Population Flows: Immigration aspects 2007–2008*, Canberra: Department of Immigration and Citizenship.

Druker, J. and White, G. (1996) *Managing People in Construction*, London: IPD.

Dryburgh, H. (1999) Work hard, play hard: women and professionalization in engineering: adapting to the culture, *Gender & Society*, 13(5), 664–82.

Duncan, R.I. and Mortimer, J. (2007) Race equality and procurement: an investigation into the impact of race equality policy on the procurement of Black and Minority Ethnic (BME) contractors and consultants in the Welsh social housing sector, *Construction Management and Economics*, 25(12), 1283–93.

EHRC (2009) *Race discrimination in the construction industry*, Inquiry Report, Manchester: Equality and Human Rights Commission.

Eliufoo, H. K. (2007) Gendered division of labour in construction sites in Zanzibar, *Women in Management Review*, 22(2), 112–21.

Ely, R. J. and Thomas, D. A. (2001) Cultural diversity at work: the effects of diversity perspectives on work group processes and outcomes, *Administrative Science Quarterly*, 46(2), 229–73.

Equalities Review (2007) *Fairness and freedom: the final report of the Equalities Review*, London: Equalities Review.

Etzkowitz, H., Kemelgor, C. and Uzi, B. (2000) *Athena Unbound: the advancement of women in science and technology*, Cambridge: Cambridge University Press.

Faulkner, W. (2006) *Genders in/of Engineering: a research report*, Edinburgh: University of Edinburgh.

Fawcett Society (2009) Are *women bearing the burden of the recession*. Available at: www.engender.org.uk/UserFiles/File/news/Arewomenbearingtheburdenofthe recession.pdf.

Fielden, S. L., Davidson, M. J., Gale, A. W. and Davey, C. L. (2000) Women in construction: the untapped resource, *Construction Management and Economics*, 18(1), 113–21.

Fielden, S. L., Davidson, M. J., Gale, A. W. and Davey, C. L. (2001) Women, equality and construction, *Journal of Management Development*, 20(4), 293–304.

Fielding, J. and Glover, J. (1999) Women science graduates in Britain: the value of secondary analysis of large-scale data sets, *Work, Employment and Society*, 13 (2), 353–67.

Foot, M. and Hook, C. (2002) *Introducing Human Resource Management*, Harlow: Pearson Education.

Fredman, S. (2002) *Future of Equality in Britain*, Manchester: Equal Opportunities Commission.

Gale, A. (1994) Women in non-traditional occupations: the construction industry, *Women in Management Review*, 9(2), 3–14.

Gale, A. and Davidson, M. (eds) (2006) *Managing Diversity and Equality in Construction: Initiatives and Practices*, Abingdon: Taylor & Francis.

Gornick, J. and Meyer, M. (2009) Gender Equality: transforming family divisions of labor, Wright, E.O. (ed.) *Real Utopias Project,* London: Verso.

Government Equalities Office. (2009) *The Economic Downturn: the concerns and experiences of women and families*, London: Government Equalities Office.

Greed, C. (2000) Women in the construction professions: achieving critical mass, *Gender, Work and Organization*, 7(3), 181–95.

Greenhaus, J. H. and Callanan, G. A. (1994) *Career Management*. Orlando, FL: Dryden Press.

Greenwood, M. (1997) *Discovering and individual's risk behaviour profile*, paper presented at ARCOM 13th Annual Conference.

Griffiths, M., Keogh, C., Moore, K., Tattershall, A. and Richardson, H. (2006) *Managing diversity or valuing diversity? Gender and the IT labour market*, Salford: Information Systems Institute.

Harel, G. H., Tsafrir, S. S. and Baruch, Y. (2003) Achieving organisational effectiveness through promotion of women in managerial positions: HRM practice focus, *International Journal of Human Resource Management*, 14(2) 247–63.

Haynes, N. and Love, P. (2004) Psychological adjustment and coping among construction project managers, *Construction Management and Economics*, 22(2), 129–40.

Henwood, F. (1996) WISE choices? Understanding occupational decision-making in a climate of equal opportunities for women in science and technology, *Gender and Education*, 8(2), 199–214.

HESA (2009) *All Students by Subject of Study, Domicile and Gender*, Higher Education Statistics Agency. Available at: www.hesa.ac.uk (accessed October 2009).

Hofstede, G. (1984) *Culture's Consequences: Comparing values, behaviours, institutions and organizations across nations*, London: Sage.

Hofstede, G. (2003) *Cultures and organizations. Software of the mind: intercultural cooperation and its importance for survival*, London: Profile Books.

Hogarth, T., Owen, D., Gambin, L., Hasluck, C., Lyonette, C. and Casey, B. (2009) *The equality impacts of the current recession*, Manchester: Equality and Human Rights Commission.

Hossain, J. B. and Kusakabe, K. (2005) Sex segregation in construction organisations in Bangladesh and Thailand, *Construction Management and Economics*, 23(6), 609–19.

HSE (2010) *Construction*, Health and Safety Executive, Available at: www.hse.gov.uk/migrantworkers/construction.htm (accessed October 2010).

Itzin, C. (1995) The gender culture, C. Itzin and Newman J. (eds), *Gender, Culture and Organizational Change: putting theory into practice*, London: Routledge.

Jackson, S. E., Brett, J. F., Sessa, V. I., Cooper, D. M., Julin, J. A. and Peyronnin, K. (1991) Some differences make a difference: individual dissimilarity and group heterogeneity as correlates of recruitment, promotions and turnover, *Journal of Applied Psychology*, 76, 675–89.

Janson, N. (1994) *Safety Culture: a study of permanent way staff at British Rail*, Amsterdam: Vrije University.

Jehn, K. A., Northcraft, G. B. and Neale, M. A. (1999) Why differences make a difference: a field study of diversity, conflict and performance in workgroups, *Administrative Science Quarterly*, 44, 741–63.

Jewson, N. and Mason, D. (1986) The theory and practice of equal opportunities policies: Liberal and radical approaches, *The Sociological Review*, 34(2), 307–34.

Jex, S. M. and Britt, T. W. (2008) *Organizational Psychology: a scientist–practitioner approach*, New Jersey: John Wiley & Sons.

Kimmel, M. S. (2009) Gender equality: not for women only M. O. Ozbilgin (eds) *Equality, Diversity and Inclusion at Work: A research companion* Edward Elgar, UK, 359–71.

Kirton. G. and Greene, A-M. (2009) The costs and opportunities of doing diversity work in mainstream organisations, *Human Resource Management Journal*, 19(2), 159–75.

Kochan, T., Bezrukova, K., Ely, R., Jackson, S., Joshi, A., Jehn, K., *et al.* (2003) The effects of diversity on business performance: report of the diversity research network, *Human Resource Management*, 42, 3–21.

Kossek, E. E. and Pichler, S. (2009) EEO and the management of diversity, P. Boxall, Purcell J. and Wright P. (eds) *The Oxford Handbook of Human Resources Management*, Oxford: Oxford University Press.

Kronsell, A. (2005) Gendered Practices in Institutions of Hegemonic Masculinity: Reflections from feminist standpoint theory, *International Feminist Journal of Politics*, 7(2), 280–98.

Kulik, C. T. and Roberson, L. (2008) Diversity initiative effectiveness: what organizations can (and cannot) expect from diversity recruitment, diversity training, and formal mentoring programs, A. P. Brief (ed.) *Diversity at Work*, Cambridge: Cambridge University Press.

Lewis, S. and Lewis, J. (eds) (1996) *The Work–Family Challenge: rethinking employment.* London: Sage.

Lewis, S. and Roper, I. (2008) Flexible working arrangements: from work–life to gender equity policies, Cartwright, S. and Copper, C. L. (eds), *The Oxford Handbook of Personnel Psychology*. Oxford: Oxford University Press.

Liff, S. and Cameron, I. (1997) Changing equality cultures to move beyond 'women's problems', *Gender, Work and Organization*, 4(1), 35–46.

Lillie, N. and Greer, I. (2007) Industrial relations, migration and neoliberal politics: the case of the European construction sector, *Politics and Society*, 35(4): 551–81.

Ling, F. Y. Y. and Poh, Y. P. (2004) Encouraging more female quantity surveying graduates to enter the construction industry in Singapore, *Women in Management Review*, 19(8), 431–36.

Lingard, H. (2003) The impact of individual and job characteristics on 'burnout' among civil engineers in Australia and the implications for employee turnover, *Construction Management and Economics* 21(1), 69–80.

Lingard, H. and Francis, V. (2006) Work–life balance in construction: promoting diversity, A. W. Gale and Davidson, M. J. (eds), *Managing Diversity in the Construction Sector*, Spon Press, London.

Lingard, H. and Saunders, A. (2004) Occupational rehabilitation in the construction industry of Victoria, *Construction Management and Economics*, 22(10), 1091–101.

Lingard, H. and Sublet, A. (2002) The impact of job and organisational demands on marital or relationship satisfaction and conflict among Australian civil engineers, *Construction Management and Economics*, 20(6), 507–21.

Loosemore, M. and Chau, D. W. (2002) Racial discrimination toward Asian operatives in the Australian construction industry, *Construction Management and Economics*, 20(1), 91–102.

Loosemore, M., Dainty, A. and Lingard, H. (2003) *Human Resource Management in Construction Projects: strategic and operational approaches*, London: Taylor & Francis.

Loosemore, M. and Phau, F. (2010) Responsible Corporate Strategy, Construction and Engineering: 'Doing the right thing?' , Abingdon: Spon Press.

Loosemore, M., Phua, F., Dunn, K. and Ozguc, U. (2010) Operatives experiences of cultural diversity on Australian construction sites, *Construction Management and Economics*, 28(2): 177–88.

Love, P. and Edwards, D. (2005) Taking the pulse of UK construction project managers' health: influence of job demands, job control and social support on psychological wellbeing, *Engineering, Construction and Architectural Management*, 12(1), 88–101.

McColgan, A. (2005) *Discrimination Law: text, cases and materials*, Oxford: Hart publishing.

McIlwee, J. S. and Robinson, J. G. (1992) *Women in Engineering: gender, power and workplace culture*, Albany, NY: State University of New York Press.

Meager, N. and Hill, D. (2005) *The Labour Market Participation and Employment of Disabled People in the UK*, working paper WP1, Brighton: Institute for Employment Studies.

Mills, A.J. (1988) Organisation, gender and culture, *Organization Studies*, 9(3), 351–69.

Mychasuk, E. (2008, 16 October) Silver-lining in recession for women, *Financial Times*.

Ness, K. (2010) '*I know Brendan; he's a good lad': the evaluation of skill in the recruitment and selection of construction workers*, Egbu, C. (ed.) Procs 26th Annual ARCOM Conference, 6–8 September 2010, Leeds, UK, Association of Researchers in Construction Management, 543–52.

Newton, R. and Ormerod, M. (2005) Do disabled people have a place in the UK construction industry?, *Construction Management and Economics*, 23(10), 1071–81.

Noon, M. (2007) The fatal flaws of diversity and the business case for ethnic minorities, *Work, Employment and Society*, 21(4), 773–84.

NSW Government (2007) *Aboriginal Participation in Construction Guidelines: applying to projects commencing 1 January 2007*, Sydney: NSW Government.

O'Reilly, C. A., Williams, K. Y. and Barsade, S. G. (1998) Group demography and innovations: does diversity help?, D. Gruenfeld, Mannix, B. and Neale M. (eds.), *Research on Managing Groups and Teams* (183–207), Stanford: JAI Press.

Pilcher, J. and Whelehan, I. (2004) *50 Key Concepts in Gender Studies*, London: Sage.

Powell, A., Bagilhole, B. and Dainty, A. (2006) The problem of women's assimilation into UK engineering cultures: can critical mass work?, *Equal Opportunities International*, 25(8), 688–99.

Powell, A., Bagilhole, B. and Dainty, A. (2010) *Achieving gender equality in the construction professions: lessons from the career decisions of women construction students in the UK*, Egbu, C. (ed.), Proceedings of the 26th Annual ARCOM Conference, 6–8 September 2010, Leeds, UK, Association of Researchers in Construction Management, 573–82.

Powell, A., Hassan, T. M., Dainty, A. R. J. and Carter, C. (2009) Note: exploring gender differences in construction research: A European perspective, *Construction Management and Economics*, 27(9): 803–07.

Pringle, J. (2009) Positioning workplace diversity: critical aspects for theory, M.O. Ozbilgin (ed.), *Equality, Diversity and Inclusion at Work: a research companion*, Edward Elgar, UK, 75–87.

Ravisi, D. and Schultz, M. (2006) Responding to organisational identity threats: exploring the role of organizational culture, *Academy of Management Journal*, 49, 433–58.

Sang, K. J .C. (2007) *The health and wellbeing of architects and the role of gender*, Unpublished PhD thesis, Loughborough: Loughborough University.

Sang, K. J. C., Dainty, A. R. J. and Ison, S.G. (2010a) *Occupational stress in the UK architectural profession: is hegemonic masculinity to blame?* Paper presented at the Equality, Diversity and Inclusion conference, Vienna, 14–16 July.

Sang, K. J. C., Caven, V. and Ozbilgin, M. (2010b) *Creating the architect: socialisation into the UK architectural profession*, paper presented at the Gender, Work and Organisation conference, Keele, 21–23 June.

Savoie, E. J. and Sheehan, M. (2001) Highlights of the conference 'diversity in the workplace: challenges and opportunities, E. J. Savoie and Sheehan M. (eds), *Diversity in the Workplace: challenges and opportunities*, Michigan: Wayne State University Press.

Schein, E. H. (1992) *Organizational Culture and Leadership: a dynamic view* (2nd edn), San Francisco: Jossey-Bass.

Schellekens, P. and Smith, J. (2004) *Language in the construction industry: communicating with second language speakers*, report for CITB-Construction Skills.

Sharp, R. Franzway, S., Mills, J. and Gill, J. 2011, Flawed Policy, Failed Politics? Challenging the sexual politics of managing diversity in engineering organizations, *Gender work and organization.* Early view, available at: doi:10.1111/j.1468-0432.2010.00545.x

Stonewall (2010) *Diversity Champions Programme,* London: Stonewall. Available at: www.stonewall.org.uk/workplace/1447.asp (accessed November 2010).

Sutherland, V. and Davidson, M. (1993) Using a stress audit: the construction manager experience in the UK, *Work and Stress,* 7(3), 273–86.

Torrington, D., Hall, L. and Taylor, S. (2008) *Human Resource Management,* Prentice Hall: Harlow.

Tsui, A. S., Egan, T. D. and O'Reilly, C. A. (1992) Being different: relational demography and organisational attachment, *Administrative Science Quarterly,* 37, 549–79.

van den Brink, Y. and Anagboso, M. (2010) Understanding the divergence between output and employment in the UK construction industry, *Economic and Labour Market Review,* 4(3), 42–51.

Wajcman, J. (1998) *Managing Like a Man: Women and men in corporate management.* Cambridge: Polity Press.

Watts, J. (2007) Porn, pride and pessimism: experiences of women working in professional construction roles, *Work, Employment and Society,* 21(2), 219–316.

Watts, J. (2009) Leaders of men: women 'managing' in construction, *Work, Employment and Society,* 23 (3), 512–30.

Whittock, M. (2002) Women's experiences of non-traditional employment: is gender equality in this area possible?, *Construction Management and Economics,* 20(5), 449–56.

Williams, F. M. and Emerson, C. J. (2001) *Feedback loops and critical mass: the flow of women in science and engineering.* Available at: http://www.mun.ca/cwse/GASAT_2001.pdf (accessed July 2008).

Williams, J. (2001) *Unbending Gender: why family and work conflict and what to do about it.* Oxford: Oxford University Press.

8 Employment relations in construction
Stewart Johnstone and Adrian Wilkinson

Employment relations and the employment relationship

This chapter explores the meaning and scope of employment relations, and how it has evolved as field of study as well as an area of management practice. It then considers how macro-level contextual factors, such as the nature of employment practices at a national level, as well meso-level sectoral contextual factors influence the regulation of work in the construction industry, and in particular the notion of the 'flexible firm'.

What is employment relations?

Employment relations (also referred to as industrial relations) can be viewed both as a field of academic study as well as a central component of HRM theory and practice. As a field of study, employment relations has been defined as 'the study of rules governing employment, together with the ways in which the rules are made and changed, interpreted and administered' (Clegg 1979). Sisson (2010, 4) defines employment relations as the study of 'the institutions involved in governing the employment relationship, the people and organisations that make and administer them, and the rule making processes that are involved, together with their economic and social outcomes'. The field of employment relations is interdisciplinary and draws upon a range of social science disciplines including economics, history, law, management studies, politics, sociology and psychology (Ackers and Wilkinson, 2003).

In terms of scope, employment relations has traditionally focussed upon exploring the collective processes, structures and actors associated with the governance of employment (Heery et.al, 2008), including collective bargaining, trade unions, employer associations, and workplace conflict. Contemporary employment relations, however, is concerned with a broader range of issues including management style and practices, work organisation, workforce development, personnel policies, and the nature of the relationships between employers and workers, irrespective of trade union presence (Sisson, 2010). Specific areas of interest typically include the underlying nature of the relations between employers and employees, the role of trade unions as representatives of employees, and the processes through which employers negotiate, consult and communicate with employees.

The broadening of the scope of employment relations is arguably reflected in the shift in terminology from 'industrial relations' to 'employment relations'. The employment relations domain has gradually expanded and become more inclusive, by exploring the nature of work in the modern service sector (as well as traditional manufacturing environments), in non-union workplaces (as well as highly unionised establishments) and in professional occupations (as well as manual or blue collar work). Contemporary employment relations is also concerned with the nature of employment in both the public and private sectors, as well as in both large and small firms. Reflecting the interdisciplinary nature of the field, employment relations concerns include the economic, societal, political and ethical aspects of employment, as well the contested and countervailing tensions which often characterise the employment relationship (Budd, 2004; Sisson, 2010).

In short, employment relations informs many of the contemporary debates around the management of people at work. The conduct of workplace relations affects all workers in some way, regardless of whether they are a senior manager in a large construction multinational, a craftsperson working for a small family-run firm, or a self-employed contractor. The rules governing work are of importance to workers as many spend a significant amount their time in work, and the nature of workplace relations affects whether this experience is positive or negative (Blyton and Turnbull, 2004). Equally, employment relations issues also concern employers seeking to harness the consent, co-operation and commitment of the workforce, making knowledge and competence in managing such issues an important component of management expertise (Doloi, 2009). The overall quality of employment relations can affect a range of issues, and in the context of construction, poor employment relations may have an impact on levels of work quality, innovation, project delays, and corporate reputation. In simple terms, employment relations issues can affect both the ability of firms to meet client needs, and the day-to-day experiences of those working in the sector. Ongoing debates in construction – such as skill shortages, health and safety, or migrant working – all have important employment relations dimensions. They also have potential implications for the competitiveness and productivity of individual construction firms, as well as the industry as a whole.

The conduct of employment relations is contingent upon a range of contextual factors at national, industry and workplace level, and which influence the regulation of work. At national level, factors such as legal regulations and cultural norms vary significantly internationally, and determine what kind of relations between firms and workers are deemed acceptable or unacceptable. Scandinavian construction workers, for example, are considered to be well paid and protected in comparison with their counterparts in China and India (ILO, 2001). Even within groups of countries which share important cultural similarities, such as Australia, the UK, and the USA, there are important institutional differences regarding the conduct of employment relations. Furthermore, within the parameters of a particular national context, employers will typically have a degree of latitude and choice regarding how they manage employment relations issues, but will also be influenced partly by industry and sectoral norms. For example, most construction work is carried out *in situ* at the project location, and the structure of the construction industry in most developed economies is characterised by the domination of major projects by a few international contractors, who in turn rely upon subcontracting (McGrath-Champ and Rosewarne, 2009). Often this requires engagement with local suppliers, and in international projects the construction industry is influenced by local employment laws, regulations and institutions (ILO, 2001).

Firms operating across international borders may adopt different employment relations approaches in different territories in an attempt to achieve a better fit with local conditions, or they may attempt to harmonise policies and transfer practices from their home country to overseas operations. Alternatively, the subcontracting business model enables international contractors to minimise employment obligations associated with direct employment, and to transfer employment relations issues and responsibilities onto domestic subcontractors. Finally, employers may be influenced by fashionable management techniques. Examples include 'Lean Production' and 'Business Process Re-engineering' which have attracted the attention of employers worldwide, though the employment relations implications have been controversial (Green and May, 2003).

The employment relationship

The employment relationship concerns the relationship which exists between employers who are buying labour to carry out work, and workers who are selling their labour and their availability to work. This relationship is complex, multi-faceted and distinctive and has four important dimensions (Wilton 2010):

- **Economic relationship**—An economic (monetary) reward is offered in exchange for the effort exerted by workers. This is often referred to as the 'wage-effort' bargain. Most people require pay in order to achieve and sustain the lifestyle they desire, and as such economic concerns are central to understanding the employment relationship.
- **Legal relationship**—Often a formal contract of employment exists which outlines the obligations of the employer and employee in relation to the exchange (e.g. hours of work, rates of pay, holidays) as well as the more

general legal obligations of the employer such as complying with health and safety laws. However, in construction a range of legal relationships exists depending on whether a person is considered in law to be a 'worker', an 'employee', or 'self-employed'. Laws determining employment status can be complex, especially in the construction industry's highly fragmented subcontracted structure. In some cases, the legal relationship between the purchaser of labour and the seller of labour can be even more ambiguous with the exchange conducted informally on the 'grey market' out of view of employment or tax legislation.

- **Political relationship**—The employment relations perspective recognises that power is often tipped in favour of the employer, and that the labour market power of individual and groups of employee varies. Individuals who possess extremely specialised skills and knowledge may command more favourable levels of power in the labour market, compared to those who possess skills which are more readily available and easily substitutable. Workers with higher levels of labour market power may find they are in a stronger position to negotiate desired terms and conditions and possess a greater degree of choice regarding the type of work they undertake. On the other hand others may feel they have very little choice over the employment they undertake. Those with low levels of individual power in the labour market may bolster their power by joining together as a collective, partly because a large group of workers withdrawing their co-operation can be more disruptive and difficult for employers to circumvent. This is the premise upon which the trade union movement was founded.

- **Psychological relationship**—The relationship between employers and employees also has many psychological aspects. Psychological concerns include the mutual expectations of the employer and the employee. In contrast to the formal legal relationship, this relationship is normally informal and unwritten and is sometimes referred to as the 'psychological contract'. The nature of this relationship has important implications as it recognises that while workers may offer their availability and skills in the labour market, this need not necessarily translate into high levels of motivation, loyalty or commitment.

Accordingly, Colling and Terry (2010) suggest that there are three important characteristics of the employment relationship, which make it distinctive from other types of contracts or relationships.

- First, the employment relationship is *indeterminate*. Employees sell their labour, which is essentially their ability to work for the employer and their availability to undertake work. This does not automatically guarantee that the employer will receive the desired levels of motivation, commitment or performance.

- Second, the employment relationship is *unequal* given that most working people are required to accept work for financial reasons. This may mean employees engage in work activities in which they have little interest or motivation. Vaid (1999) for example found that 90% of construction workers in India reported that they believed they had no choice but to undertake construction work. Employers on the other hand, generally have more choice regarding HRM decisions such as whom to employ, how to organise work, or what levels of training to provide. Of course, there are always parameters which constrain management choices such as the nature of product, industry and labour markets. Recurrent labour shortages in the construction labour markets and a lack of local construction workers in countries including the UK, Australia, USA, Malaysia and Canada illustrate how employers do not necessarily have a free choice over available labour (Dainty *et al.*, 2005; McKenzie, *et al.*, 2000).
- Third, in common with product and labour markets, the employment relationship is *dynamic*. This means that while at particular point in time the employer may enjoy a power advantage, the balance and nature of the power differential inevitably shifts over time with economic, political, social, cultural and demographic trends.

The employment relationship in the 'flexible firm'

By the nineteenth century a 'standard' model of employment had emerged (Winch, 1998), characterised by a 'direct employment' model where employees worked for a single employer, and often for the same employer for the majority of their careers. In many sectors, the employment relationship was relatively stable in terms of legal, economic and psychological dimensions. However, the construction industry is dominated by project-based work of a determinate duration, and economic and contractual factors have long exerted a strong influence over dominant personnel strategies (Druker *et al.*, 1996). Extensive labour outsourcing has been the norm in most countries (ILO, 2001), and this has made relationships between employers and employee more fluid, transient and opaque than in many other sectors. The highly cyclical nature of the demand for construction projects means maintaining a flexible labour force which can expand and contract to meet peaks and troughs in demand has been a key concern in the sector, resulting in most firms staffing for troughs rather than peaks (Green and May, 2005; McGrath-Champ and Rosewarne, 2009). Equally, the wide range of tasks required on a project can vary considerably, meaning the skills and expertise required are also highly variable, even at different points on a single project (Dainty *et al.* , 2006).

The desire to maintain flexibility has meant that subcontracting and outsourcing arrangements have been central to the business model of the industry worldwide for some time (Druker *et al.*, 1996; Forde *et al.*, 2009).

Druker and Croucher (2000) suggest use of casual contracts and subcontracting has become even more widespread in the last few decades across several European nations including Germany, Sweden, the Netherlands and Spain. As a result, while full-time permanent employment remains the norm in many countries and industrial sectors (Nolan and Wood, 2003), in construction 'non-standard' or 'atypical' models of employment, such as fixed-term work and self-employment have a long genesis (Beardsworth *et al.*, 1987). For example, Raidén *et al.* (2007) estimate that around 36 per cent of the UK construction workforce is self-employed, compared to 12 per cent of the UK workforce. Winch (1998) also notes that the extensiveness of casualised forms of labour, and labour-only subcontracting in particular, has been much greater in the UK than in several other European nations (Winch, 1998).

With construction projects, the workforce delivering a construction project is often highly fragmented, with the composition of labour changing on a project-by-project and even day-to-day basis (Dainty *et al.*, 2006). The sector differs from many others in terms of the extensive use of subcontracting, joint ventures, alliances, and even the creation of new organisations to deliver a particular project (Rubery *et al.*, 2004). This complexity has potentially far-reaching implications for wider human resource and employment relations issues including pay and reward, career development and employee representation. HR policies can become 'fragmented, disjointed and blurred', making the employment relationship more complex than many other sectors where the boundaries are more explicit (Marchington *et al.*, 2010, 471). As Rubery *et al.*, note (2004,1):

> The notion of a clearly defined employment relationship becomes difficult to uphold under conditions where the employee is working in project teams, or on a site alongside employees from other organisations, where responsibilities for performance and for health and safety are not clearly defined, or involve organisations other than the employer (Rubery, *et al.*, 2004, 1).

As highlighted earlier, a variety of commercial and employment relationships may exist, each with distinctive implications. While subcontracting allows contractors to effectively free themselves of some direct employment obligations, it requires that contractual relationships with subcontractors are well managed, and assumes that in turn they will manage and motivate their workforce effectively (Druker and White, 1996).

Theoretical perspectives

Theoretically, issues of organisational and workforce flexibility can be thought of in terms of an influential model of flexibility developed in the 1980s by Atkinson (Atkinson, 1984, 1985; Atkinson and Meager, 1986).

Four forms of flexibility are identified: numerical; functional; distancing and pay.

- **Numerical flexibility**—refers to the 'ability to adjust the level of labour inputs to meet fluctuations in output' (Atkinson and Meager, 1986, 3). In other words, the degree of flexibility over the number of people employed or the actual hours worked. Given peaks and troughs in demand for projects, maintaining numerical flexibility of labour levels has always been a priority in construction.
- **Functional flexibility**—refers to 'the firm's ability to adjust and deploy the skills of its employees to match the tasks required by changing workload, product method or technology' (Atkinson and Meager, 1986, 4). It is therefore concerned with developing multi-skilled employees who can work flexibly across a variety of tasks in line with changing business requirements, as opposed to a very narrow specialisation.
- **Distancing**—refers to the use of subcontracting and outsourcing arrangements to minimise the amount of activity deemed 'non-core' carried out by the employer, based on the view that it could be carried out by an external organisation more effectively. While in many sectors tasks commonly outsourced include cleaning, catering and security, construction organisations have generally had a much broader view of what constitutes core activity for a long time. It is therefore common for a large construction contractor to retain a relatively small core workforce, with a high proportion of labour provided through subcontracting arrangements, often by small entrepreneurial firms, agencies, and self-employed personnel.
- **Pay flexibility**—refers to 'the extent to which a company's pay and reward structure supports and reinforces the various types of numerical and/or functional flexibility sought' (Atkinson and Meager, 1986, 4). An example of this could be by linking pay to performance of individuals, teams, or the enterprise.

According to Atkinson's model workers can be divided into two main groups: 'core' workers and 'peripheral' workers. Core workers are likely to be highly regarded permanent employees, well paid, and enjoy good career prospects, though in return they are expected to be functionally flexible. They will be involved in key business activities requiring high levels of skill and expected to be adaptable across a range of tasks. Examples of such roles may include professional engineers, and commercial, building and construction management roles. Peripheral workers, on the other hand, provide businesses with a high level of numerical flexibility, in terms of both the number of staff working at a particular time, as well as over the amount of hours worked. They have less job security and may be employed on short or fixed-term contracts or supplied by an employment agency to complete a specific job. Examples of such roles may include a range of skilled and semi-skilled personnel in areas including craftworkers, operatives, labourers and administrators, as well as specialist professional staff. Employees in this category are likely to have significantly less responsibility and discretion over the work undertaken. The population of peripheral workers includes those who are self-employed, those selling their labour only (labour-only sub-contractors), as well as those offering both labour, material and equipment and trading as small businesses.

A flexible workforce can have various advantages and disadvantages for employers and workers. For employers, commonly cited benefits include the ability to respond to cyclical and unpredictable demand by providing access to the required skills and to vary labour levels on an *ad hoc* basis. Limited numbers of direct employees also allow contractors to effectively outsource many of the labour and employment relations issues to third parties. However, given that those working together on a project may be employed on different sets of terms and conditions, for different periods of time and with contractual obligations to different employers, a potential challenge concerns the requirements of effective project co-ordination and employee motivation (Atkinson, 1984). This leads to challenges in terms of work quality, cost escalations and project completion delays. For workers, working for a range of employers on projects might offer a stimulating variety of work and the opportunity to have some choice over where and when to work. On the other hand, fragmentation, casualisation and individualisation have led to many concerns being raised regarding the work experiences of contingent workers. These include access to adequate training, development and career opportunities, job insecurity, health and safety risks, and a lack of access to employee representation or social security (Druker and White, 1996; Green and May, 2003).

Case Study 8.1

Managing the flexible workforce at BuildCo

BuildCo is the civil engineering division of a major international contracting organisation, operating autonomously as one of several divisions within the highly devolved management structure of the parent organisation. BuildCo is highly regarded in the industry and has a strong track record of winning prestigious major projects, ranging from construct-only works contracts to more complex 'asset-based service driven solutions'. Typical projects include highways, marine, tunnelling, water supply and rail infrastructure. BuildCo employs over 1,500 staff, primarily in a range of managerial, administrative and technical functions.

Reflecting the emphasis upon subcontracting, tight cost control and responsiveness to demand cycles, the approach to labour management is strongly influenced by the notion of the 'flexible firm'. In order to deliver the wide range of activities required by a project, BuildCo typically engages specialist companies to undertake specific work tasks for defined periods of time, and also 'buys in' general operative, production and craft labour sourced through employment agencies. Indeed, much of the actual execution of a project is conducted by external suppliers, with BuildCo focusing upon project bidding, conception, scheduling, and management activities. BuildCo therefore has three distinctive groups of staff. Firstly, there is a core management team responsible for commercial issues, with responsibility for bidding for work and driving the strategic agenda of the business. Secondly, there are construction supervisors directly employed by BuildCo with responsibility for day-to-day project management at an operational level. The third group of staff consists of white and blue-collar external labour which is utilised and bought in on a panoply of flexible contractual arrangements. The flexible workforce model was believed to offer several benefits to the business. Most importantly was the ability to expand and contract the size of workforce in response to economic cycles and demand for projects, which meant the business could align labour utilisation with workflows more efficiently. It also allowed access to specialist skills it might be difficult or undesirable to maintain in-house. Though some relationships with sub-contractors would be on a one-off basis, BuildCo also engaged with a range of providers and preferred to develop close and enduring relationships with tried and tested suppliers where possible.

However, there are also disadvantages to this business model. For example, with regard to the peripheral workforce, it was recognised that it can be difficult to engender a desired corporate culture of quality, service and safety among those working as part of the transient peripheral workforce, often working at some distance from the main contractor. This was believed to be challenging given the business emphasis on driving operational efficiencies, improving productivity and eliminating waste. It was also believed to be counter to the desire to re-use knowledge and learning from project-to-project. There were also challenges gaining the loyalty of 'core' employees and this was manifested in high levels of turnover among the directly employed professional workforce. This was partly attributed to a lack of career development opportunities within the division, and practices which encouraged specialisation among core workers rather than the development of a broader portfolio of skills and expertise. Internal redeployment of transfers within the parent company were also rare. Recognising the challenges of staff turnover, BuildCo has begun to place greater emphasis upon creating clearer and more structured development and career pathways for its core workforce from graduate recruitment through to senior managerial and commercial roles.

Conflict and co-operation: theory and practice

Managing the tensions between conflict and co-operation is a key concern of employment relations (Edwards, 2003). Within the construction industry, the multiplicity of parties and workers involved in the delivery of a project means there are arguably even more challenges associated with managing a diverse range of interests. In this section we consider some of these tensions and how they compare across different national contexts. The influential sociological concept of 'frames of reference' is discussed (Fox, 1966), and a case study illustrates the relevance of these issues to employment relations in the construction context.

Managing convergent and divergent interests

Employers and employees have both convergent and divergent interests, and as such are involved in issues which may elicit both conflict and co-operation (Edwards, 2003). Employers and employees may well have shared interests, such as the future success of the organisation, but for very different reasons. For employers, a key driver might be to improve competitiveness in order to increase profits and meet the expectations of shareholders and investors. For employees, interest in future business success may be because a successful and sustainable organisation will offer greater job security, a more positive working environment and better career opportunities than a poorly performing organisation. The compliance and co-operation of workers with management objectives is not automatic, irrespective of whether or not a formal and direct contractual relationship exists. Employers and workers may hold very different views regarding the implications of particular business decisions. Crucially, and unlike other resources such as technology, workers are not passive beings but have the ability to question, challenge and resist management demands. Employees can also exercise a degree of discretion regarding what kind of employment to accept and the amount of effort they exert. Sometimes employees will be motivated to work to the best of their ability, or to even 'go the extra mile' (often referred to as 'discretionary effort') in order to meet a project deadline. However, such levels of commitment can be extremely difficult for employers to capture as it requires both employee consent and engagement.

Where workers are unhappy and dissatisfied with their work, resistance may be noticeable as employees attempt to reassert a level of control over their work. Manifestations of resistance include withdrawing both physically and psychologically from the work they are paid to do. Examples of physical withdrawal include poor timekeeping, absenteeism or even resignation. Psychological withdrawal might be characterised by employees working very slowly or in a robotic manner and exerting the minimum amount of effort required to carry out their role. In addition to discontent and dissatisfaction among individual employees, a more visible manifestation of discontent is when it occurs at a collective level culminating in industrial action such as strikes (Hyman, 1975). Conflict may occur as employees make a decision to act collectively in order to promote their interests, achieve particular objectives or express high levels of dissatisfaction with an employer. Given the complex nature of the concerns around the regulation of the wage-effort bargain and the rules and policies which govern work, employment relations has often been concerned with what has been termed the 'institutionalisation' of conflict (Flanders, 1965).

Recent data suggests that in the construction industry the most strike-prone nations include Denmark, Finland, Spain and Canada. Conversely, countries with a relatively low incidence of disputes leading to working days lost include Austria, Sweden, Germany, Malta, US, Japan and the UK (Hale, 2008). In the case of the UK, levels of strike activity have generally declined since the 1970s, with the last large-scale national strike in the construction industry now 40 years ago. While strike activity has dissipated in many advanced industrial economies, this does not mean workplace conflict has disappeared altogether. In the UK, unofficial industrial action occurred at the Lindsey Oil Refinery in 2009 (see Case Study below). Even in Dubai, where there are no formal rights to join a trade union or to take industrial action, work stoppages involving 2,500 labourers interrupted construction during the prestigious Burj Tower projects of 2006. This highlights how even in nations where formal labour rights are minimal, and the construction labour market is fluid with a plentiful supply of low cost migrant labour, the potential for conflict does not disappear. This underlines the continued importance of considering employment relations issues.

Case Study 8.2

Unofficial industrial action at the Lindsey oil refinery (UK 2009)

In January 2009, employment relations were under pressure at the Lindsey Oil Refinery in Lincolnshire, England. The dispute involved a controversial decision by the owner of the refinery, French company Total, to award a contract to an American firm, who in turn operated a tendering process and subsequently sub-contracted the work to an Italian organisation. However, a controversial term of the Italian contract was that they would use their existing supply of Italian and Portuguese workers to undertake the work. This was deeply unpopular with both British workers and trade unions who expressed concerns regarding important projects being subcontracted to overseas firms which employee foreign workers, especially if suitable qualified labour is available locally. This was exacerbated by the fact the UK was experiencing a deep economic recession resulting in a general climate of job insecurity and rising unemployment across the economy. The result was unofficial strike action at the plant, as well as various other protests and demonstrations of support at other oil refineries and power plants across the UK. The employer's position was that no Total employees would lose their job as a result of the contracting decision, and that local labour had been used in the early stages of the project. However, the subcontractor preferred to use their own staff. The dispute was later resolved when the contractor agreed to hire 100 'locally-sourced' workers. The case illustrates the political, economic and social aspects of employment relations. Freedom of movement across a diverse geographic area such as the European Union means such issues are likely to remain very important in the future (*Source*: *BBC News*, 2009).

Theoretical perspectives

In order to improve understanding of the nature of the employment relationship it is useful to consider how we view the world of work and organisations. Fox (1996) proposed 'frames of reference' which capture 'management ideologies' and determine how we expect people to behave, how we react to peoples' behaviour, and the methods we choose when we attempt to influence others and change peoples' behaviour. Two particularly influential frames of reference were outlined: 'unitarist' and 'pluralist'.

A *unitarist* frame of reference stresses the common goal of an enterprise, with all participants sharing the same basic aim, namely that of creating and contributing to the aim of an efficient and competitive enterprise. All participants are expected to share the rewards accrued by achieving this aim, and the organisation is viewed very much as a united team or family, without factions or rival leaders. A unitarist worldview perceives efficient organisations to be rational and machine-like in pursuit of predefined targets and objectives, and emphasises the importance of strong leadership and a commitment to teamwork (Green and May, 2003). Such a unitarist perspective suggests that conflict is not an inherent or inevitable feature of the employment relationship. Conflict is therefore viewed as a result of a lack of communication and poor management, or even as a result of troublemakers. Unitarist perspectives resonate with many of the ideas of the 'human relations school' of the early twentieth century, which highlighted a distinction between extrinsic and intrinsic rewards. Human relations' theorists suggested that workers are not just motivated by money (extrinsic reward) but also by the positive feelings and satisfaction associated with doing a job well (intrinsic reward), and as such conflict can be viewed as the result of poor social relations in the workplace. However, these views still largely overlook the potential for fundamental differences between employer and employee perspectives to arise or to result in conflict or resistance. Key criticisms include the failure to explain the prevalence of conflict in many organisations or to take into account the uneven distribution of power between management and workers, or between actors in the supply chain. While many management initiatives, including business process re-engineering and total quality management often have unitarist undertones, critics have questioned the assumption that employers and employees always share common goals, values and priorities.

A *pluralist* frame of reference, on the other hand, adopts a rather different approach, viewing organisations as coalitions of people with a variety of different interests, aspirations, perceptions, priorities and aims. Rather than the 'team' analogy of unitarism, organisations are viewed as *'a miniature democratic state composed of sectional groups with divergent interests'* (Fox, 1966, 2). As well as customers, stakeholders include shareholders, employees, trade associations, unions, suppliers and various other interest groups (Green, 1999). The view suggests that different groups of employees may possess different points of view, and differences in interests and priorities are believed to be completely natural in organisations. From a pluralist perspective, construction organisations and projects consisting of individuals from a range of backgrounds, occupations and organisations, represent a coalition of people with both converging and diverging goals. Occasional disagreements and conflicts are viewed as a normal and inevitable aspect of work and organisational life, and the aim is therefore to keep conflict within some kind of accepted bounds. Accordingly, the pluralist perspective believes that the employment relationship is inherently contested and characterised by various tensions, meaning some regulation is necessary. Examples include systems of employee and employer representation, or state regulation. While in theory, unitarist and pluralist frames of reference are polar opposites, in reality the perspectives can coexist, as organisational actors may hold complex and ever-changing ideas and views at different times, or even in different situations. For example, while managers and employees may be united around the belief that the organisation should be successful, competitive and striving to improve, they could hold very different views regarding how this might best be achieved.

Stakeholders in employment relations

The dynamics of employment relations depend partly upon the role of actors including management, the state and trade unions. The role of each in shaping employment relations can vary significantly depending upon both national and sectoral context. Relevant factors include the formality and general approach to HR management by employers, and in particular how employers view the role of government and unions in influencing workplace relations. The degree of intervention of the state and the role of trade unions also varies both by national context and by industry sector. However, in many countries the limited role of construction unions requires a different role for management and the state in developing and sustaining positive employment relations.

Management

Beardwell and Holden (1999) suggest that the role of management in the employment relationship is threefold. First, to control and direct the employees being managed, either directly by supervising or inspecting work, or indirectly through processes and procedures which need to be followed. Second, management is concerned with the co-ordination of different human and non-human activities. In a construction project environment, the need for someone to pull together the work of different teams and functional specialities is crucial, and often the primary activity of large contractors. Third, management may be concerned with motivating staff and encouraging them to comply with what management seeks to achieve. In large organisations, responsibility for employment relations strategy and policy may be part of a specialist Human Resources or Employee Relations function, though the status and relevance of the function has often been questioned in the sector (Clarke and Hermann, 2007; Druker *et al.*, 1996; Hillebrandt and Cannon, 1990). In the construction context the preponderance of small firms means employment relations issues may be an implicit part of a more general managerial role.

Earlier in the chapter we noted how employees sell their ability to be productive but that employee productivity is not necessarily automatically forthcoming. The social nature of workplace relations plays a part in determining the extent to which potential employee contribution translates into actual effort and output. In addition, management often has a degree of latitude in how it attempts to manage workplace relations, and is not entirely the victim of external forces. An example of this might be the extent to which construction firms have adopted the 'hollowed-out flexible firm' model. While the model remains dominant in some contexts, there is also evidence of some firms which have engaged much more in a direct employment model. This would suggest that the emphasis on flexibility in the sector is not merely an inevitable consequence of employment in project-based industries, but perhaps also the relationship between the underlying business model and HRM strategy. If the business model is premised upon low cost, then a flexible firm model could be viewed as a way of keeping employment and HR responsibilities at a minimum. A business model premised upon overall quality and value, however, might view a subcontracting model as problematic because of the potential management difficulties, and the challenges of achieving the buy-in, commitment and engagement of all workers believed to underpin a quality output. Employers may also be influenced by the work of professional, industry and policy bodies, such as Constructing Excellence in the UK, or the European Construction Industry Federation (ECIF), which represents member federations across 29 countries.

Various typologies have emerged in human resource management which attempt to capture the range of approaches employers may adopt in their attempt to manage the employment relationship (Guest, 1995; Purcell and Sisson, 1983; Purcell and Ahlstrand, 1993). The framework devised by Guest (1995) highlights various options and choices available to managers across two dimensions (see **Table 8.1**):

Table 8.1 Guest, D. Management Styles Typology (1995)

		HRM Priority	
		High	*Low*
Industrial relations policy	**High**	the New Realism	traditional collectivism
	Low	Individualised HRM	'the black hole'

- **The New Realism**—The commitment of employees is believed to be essential. In order to do this, there is an emphasis upon both progressive HRM practices at an individual employee level as well as working with trade unions at a collective level. From this perspective, there is no need to make a choice between HRM and trade unions; the two are viewed as complementary.
- **Individualized HRM**—A more 'individualised' employment relationship which focuses upon linking the performance of individuals to the performance of the organisation as a whole. Individual HRM practices are deemed important, but employee representative mechanisms at a collective level are not believed to be essential to engendering employee commitment. This approach is often associated with North American approaches to modern human resource management which became popular in the 1980s, but is less common in countries in Europe and Australasia where there are stronger histories of collective employee representation. Nevertheless, the continued decline in many advanced economies of collective union representation means the focus on an Individualized HRM approach is expanding. Individualised HRM is an option for employers in nations where the construction industry is lightly unionised.
- **Traditional Collectivism**—Focus on managing the employment relationship through systems of collective employee representation. Collective employment relations structures and processes are believed to be the most effective way of dealing with issues of employee representation, health and safety or discipline and grievance situations. This is generally associated with workplaces with a long history of trade union recognition and high union membership, though in many industries and nations such workplaces are in decline.
- **'The Black Hole'**—Referring to a situation characterised by the absence of trade unions as well as of progressive human resource practices. In this situation employees may be treated in unfair or arbitrary ways,

exposed to health and safety risks, and levels of employee commitment may be low. Labour is viewed as a variable cost and the emphasis may be upon minimising expenditure on labour. It may be a deliberate management choice, or the result of a lack of knowledge or interest in the importance of HRM and employment relations issues.

The choices employers make regarding the management of employment relations depends partly upon whether management perceives the employment relationship as being characterised by primarily convergent or divergent interests, as well as how managers perceive their 'right to manage'. For example, an employer might wish to improve productivity, introduce new technology, or modify ways of working. All have potentially important employment relations implications in terms of the extent to which employees are willing to accept change. Managers who hold strong views about their right to make decisions may attempt to take unilateral decisions without informing or consulting workers; others may take a more pragmatic view that in order to elicit employee co-operation and commitment, employee involvement in change is required. Managers may also have different preferences regarding the most desirable process for involving and consulting employees, and in particular whether collective representation through independent trade unions or more individual mechanisms are thought to be appropriate. Though management styles vary between firms and across countries, thus making it difficult to generalise, the construction industry has been plagued by a poor image in relation to its approach to human resource management and workforce relations (ILO, 2001), often akin to a 'Black Hole' or 'Hard HRM' style of management, or what Green (2002, 2) describes in the UK context as 'an institutionalised regressive approach to HRM'.

The state

Another important actor with an influence in the conduct of employment relations is the government. Governments may exert influence over employment issues in several ways including:

• **Setting employment laws**—Determining the rights of employees, both individually and collectively. This can include setting the basic rules regarding how employees can expect to be treated by their employer. In terms of individual employees this might include issues such as the minimum wage workers can be paid, the maximum amount of hours they can be expected to work, as well as issues of unfair treatment or discrimination. At a collective level the state may set policies regarding the rights of employees to join a trade union. Companies operating in the European Union are increasingly required to observe agreements and policies which apply across the entire region, as well as the requirements of particular national contexts. As well as general legal obligations, issues which are particularly pertinent to the construction industry include health and safety, immigration, and the use of contingent labour.

- **Economic policy**—This includes policies in relation to the nature of the welfare system, benefits, taxation, regional development initiatives. For example, tax policies which incentivise the use of self-employed labour can influence construction firms' resourcing policies in the direction of labour flexibility, and in particular reliance upon large pools of casual, self-employed and short-term workers. While labour flexibility is typically attributed to cyclical demand for projects, other issues include potential savings often associated with using contracted labour associated with National Insurance contributions, holiday pay and sickleave (Druker and White, 1996), as well as training and development costs. Government policies and tax systems can therefore influence employer and industry behaviour (McKenzie *et al.*, 2010).
- **As an employer**—In many countries public sector jobs constitute a significant component of the economy, and therefore the government can also act as a role model and exemplar of good employment practice. Governments also act as an important purchaser of construction projects, and may also stipulate specific conditions regarding the procurement of labour on public sector construction projects. However, trends towards outsourcing in the public sector mean direct employment of construction workers in the public sector has declined in many countries (ILO, 2001).

The role of the state in shaping employment relations is very much linked to wider political, social, economic, cultural and institutional forces. In some countries such as the UK and Ireland, the role of the state in the management of employment relations is often described as 'voluntarist', with a general view that minimal state intervention is desirable beyond setting some basic parameters. In such circumstances, governments take the view that the rules and regulations of employment should normally be agreed between employers, employees, and trade unions. On the other hand, in several other European countries such as Norway, Sweden, Germany and Austria, governments have adopted a more 'corporatist' approach, where the state, employers' associations (collectively representing views of employers) and trade union bodies (collectively representing views of employees) take a more active role in regulating the rules of employment. Yet even within a voluntarist framework, governments are seldom neutral; certain policies and practices may still be encouraged or discouraged both implicitly and explicitly by the state. Examples might include specific policy interventions to address skill shortages, or broader policy statements such as Royal Commissions in Australia, and the Egan Report *Rethinking Construction* (DETR, 1998) in the UK. The role of the state can also vary over time especially when different governments hold different ideological perspectives regarding economic policy (Green and May, 2003).

Trade unions

In many countries, trade unions have also been important actors in employment relations. A trade union has been defined as:

> a continuous association of wage earners for the purpose of maintaining or improving the conditions of their employment" (Webb and Webb, 1902, 1).

> an organisation (whether temporary or permanent), which consists wholly or mainly of workers ... whose principle purposes include the regulation of relations between workers ... and employers or employers' associations (Trade Union and Labour Relations (Consolidation) Act 1992, Part I).

Trade unions emerged as bodies to collectively represent the interests of workers (often skilled craft workers, before spreading more widely), who often found themselves in a subordinate role as waged labour in the employment relationship. While individual employees may have possessed limited power, by acting collectively to defend their interests as the sellers of labour, workers were able to yield more power and strengthen their position in the labour market (Ackers *et al.*, 1996). The core mechanism for advancing employee interests was the negotiation process between employers and trade unions referred to as 'collective bargaining'. This involves negotiations around the 'rules of the game'. These rules can be either 'substantive' or 'procedural'.

- **Substantive rules**—These refer to rules regarding the substance of the effort-wage exchange such as rates of pay, working hours, number of holidays
- **Procedural rules**—These refer to the rules regarding the processes and conduct of employment, such as how to settle disputes or disagreements and grievances.

The precise nature of the bargaining processes can vary considerably in terms of:

- **Level**—Are agreements negotiated nationally, locally or at site/workplace level?
- **Form**—Are negotiations formal or informal?
- **Scope**—Which items are actually subject to bargaining, and are any issues excluded?

However, union membership and trade union density (the percentage of employees who are trade union members) has declined in many countries internationally over the last few decades. In the UK, density has declined from a peak of 56 per cent in 1979 (Waddington, 1992), to 27 per cent in 2010 (15% in construction) (BIS, 2011). Union decline has also occurred in the United States, from a peak of around 35 per cent membership in the 1950s (Katz and Colvin, 2011), to 12 per cent of all workers in 2010 (13% in construction)(BLS, 2011). Decline has been more moderate in Canada, with around 30 per cent of workers members of unions, and a similar proportion of workers in the construction sector who are union members, making the sector one of the mostly highly unionised outwith the public sector. (Uppal, 2011). In Australia, construction and manufacturing industries have traditionally been the most densely unionised, compared to the more lightly unionised service sector. However, union density has fallen from a peak of 65 per cent in 1953, to just under 50 per cent in 1990, and falling to 18 per cent in 2010. Currently union density in the construction sector is around 17 per cent (Ausstats, 2011). Various explanations for the decline in union membership have been given including changing state and government policies, the globalisation of economic competition, and the changing attitudes of employers and workers regarding the relevance or effectiveness of unions. Other issues include the inability of unions to adapt to meet the changing context of work, in terms of issues such as recruitment of new members, meeting member expectations, and the effectiveness of union leadership, strategies and practices. As discussed earlier, the fragmentation, casualisation and individualisation of construction workforces, as well as the preponderance of small firms at operative and site level, arguably provide a challenging environment for collective employee representation.

Current debates in employment relations

The contemporary employment relations field is characterised by a range of important debates around the role of trade unions, employee voice and engagement, all of which have resonances for the construction industry. The general decline of trade unions raises important questions regarding their future prospects. Another topical employment relations issue concerns the perennial challenges of fostering positive, collaborative workplaces relations, with or without unions.

The future of trade unions

Attempts at revitalising trade union membership can be evaluated in terms of two main themes: (i) developing union and management partnerships and (ii) union organising.

Firstly, the notion of union-management partnership can be linked to the ideas associated with the 'New Realism' mentioned earlier in the chapter (Guest, 1995). It refers to attempts to forge more collaborative relationships between employers, employees and trade unions resulting in mutual gains which are acceptable to all parties. It is based upon the assumption that there is both a business and ethical case for unions and employers to work together, especially around issues of strategic organisational change (Stuart and Martinez-Lucio, 2005). While notions of joint working are of course not new, union management partnership can be viewed as partly inspired by the nature of union/management relationships in continental European 'social partnerships' which draw heavily on a pluralist view of the employment relationship. Geographically, social partnership can be traced to countries including Austria, Belgium, the Netherlands Germany and Switzerland (Ferner and Hyman, 1998). While partnership does not necessarily always equate with harmonious conflict-free relationships between employers and unions (Hyman, 2004), it has an emphasis on working together to minimise conflict but also dealing with it when it occurs. In addition, an influential literature has also emerged from the US which has proposed that unions can actually contribute positively to management decision-making in terms of innovation, quality and productivity (Rubenstein and Kochan, 2001) and that competitive organisations and positive HRM requires the support of multiple stakeholders in an organisation including unions (Kochan and Osterman, 1994, Lewin 2010). A recent example of such an approach to the management of employment relations in construction was the Major Projects Agreements implemented during the construction of London Heathrow Airport Terminal 5 Project undertaken in the period 2001 to 2008. The framework outlined commitments regarding pay and conditions including levels of direct employment, agency working, overtime, health and safety, and diversity and equality. The agreement also included a role for the trade unions as partners in a range of strategic, organisational and employment issues, and the approach is believed to have contributed to a range of positive performance outcomes including completion on time and budget, minimal time lost to employment relations disputes, and an above-average health and safety record (Deakin and Koukiadaki, 2009).

Secondly, union 'organising' has also emerged as a potential union revitalisation strategy, and has received attention in the USA, UK and Australia. The term can actually be used to refer to various related strategies including recruiting non-members in workplaces where unions are recognised, recruiting workers in workplaces where there is no union recognition, or an approach which emphasises the involvement and engagement of ordinary members (Simms and Charlwood, 2010). In this respect 'organising' has been positioned by some as an alternative to the partnership approach outlined earlier (Heery, 2002). The organising model has been promoted for the last decade by the Australian Council of Trade Unions (ACTU). A recent example is an effort by Australia's Construction Forestry Mining and Energy Union (CFMEU) to organise construction workers in Sydney through site visits by union organisers, and especially young workers and non-English native speakers (Leung *et al.*, 2010). However, there is mixed evidence regarding the effectiveness of organising attempts, with several studies suggesting that the success of such approaches has so far been limited (Gall, 2007). Potential challenges which may be encountered by trade unions in pursuing such strategies include high costs as well as the possibility of intense employer resistance.

Managing without unions

Employment relations in non-union firms, where trade unions are not recognised by employers for collective bargaining or employee representation, can be expected to range across an entire spectrum of employment relations arrangements. These can range from high-trust, positive employment relations, through to exploitative conditions characterised by mutual distrust and suspicion, which have been termed 'bleak house' (Sisson, 1993) or 'black hole' scenarios (Guest and Conway, 1999). While typologies have certain limitations, implying that relationships are static and clear rather than dynamic and fuzzy (Dundon, 2002), they nevertheless provide a useful way of exploring the range of options that may exist. For example, Guest and Hoque (1993) devised a typology around 'good, bad, ugly and lucky' approaches to employment relations in non-union firms.

* **Good**—Clear HRM/ER strategy, high-involvement model, training, above-average pay.
* **Bad**—No HRM/ER strategy, limited uptake of HRM practices but without malicious intent. These may be small firms with limited power in the supply chain, dependent upon large organisations for work, and an emphasis upon minimising (labour) costs in highly competitive markets.
* **Ugly**—Clear strategy, little use of HRM/ER practices and evidence to suggest management seeking to exploit workers. Again, these may be small firms seeking to minimise labour costs and maximise flexibility.
* **Lucky**—No clear HRM/ER strategy but with some innovative HRM practices.

As union membership has declined, many employers across various industrial sectors have introduced alternative employee participation mechanisms, and often these have been instigated by management. This may reflect a belief that there is a business (efficiency) case for involving employees, or an ideological (ethical) belief that involving and consulting workers is the right thing to do, though in reality, these underlying beliefs are unlikely to be mutually exclusive.

There are also various forms of employer-led participation techniques can take (Wilkinson and Dundon, 2009). First, there are communication techniques where management shares information with employees. Traditionally, this may have taken the form of memos and letters, but technological advancements mean there are now many ways employers can communicate with employees such as intranets, social media and online discussion boards. Second, there are techniques which are more concerned with generating feedback and input from the workforce, and these can be referred to as upward problem-solving techniques. Specific methods can include suggestion schemes, focus groups, quality circles and employee attitude surveys. Thirdly, participation can be focused around task-based involvement rather than managerial processes. Examples could include giving employees a wider range of tasks or additional levels of responsibility. This technique has links to notions of organisational flexibility. Finally, teamworking techniques have also attracted significant management interest, and typically involve setting up self-managed teams and devolving responsibility for some decision-making. Such teams are given the responsibility for the completion of an overall task, but often have greater discretion over how work is organised and work without direct supervision. Management-led employee involvement techniques might be created as an alternative to representative participation mechanisms such as trade unions, or may operate in parallel to such structures (Wilkinson and Fay 2011). Employers may also create their own in-house representative body such as the Balfour Beatty Group Staff Association. While formal participation techniques are now well-established in mainstream HR theory and practice, their usage is arguably more complicated in the fragmented, informal and transient construction project environment. However, in a study of the electrical and mechanical trades in construction, Coffey and Langford (1998) reveal a strong appetite for direct participation among workers but report a low uptake of such practices. They also suggest that participative work processes do occur in construction although often as part of informal and embedded working practices as opposed to a formal HRM initiative. Finally, they found little enthusiasm for indirect/representative participation.

In terms of attitudes to trade unions among employers who do not recognise trade unions, Beaumont (1995, 172) suggests two different approaches: union 'suppressionists' and union 'substitutionists'. A union substitution approach, on the one hand, is associated with a sophisticated approach to HRM issues, by for example offering highly attractive employment packages to mitigate interest in trade union activities by making them 'unnecessary'. Such employers may offer an attractive employment package which may mean that union membership is deemed less necessary by employees. Union suppression, on the other hand, involves minimising or eliminating trade union activity within organisations, and where recruitment and dismissal practices might be used partly to minimise likelihood of unionisation. An extreme example of union suppression would be 'blacklisting', the practice of screening potential workers to avoid hiring those believed to be 'troublemakers' often associated with previous union activity. A recent investigation in the UK revealed that many high-profile international contractors had performed trade union checks on workers through a private investigation agency. New laws have since been introduced in response to blacklisting activities, primarily to offer protection against loss as a result of blacklisting, but such measures have been criticised for failing to make blacklisting unlawful or to make it an offence (Ewing, 2010).

Employee voice

Increasingly management is thinking beyond a union/non-union approach and looking at ways of getting closer to staff. The expression 'employee voice' has been increasingly used in HRM in recent years. In general the term refers to how employees are able to have a say over work activities and organisational decision-making issues within the organisation in which they work. Voice is a necessary precursor for participation, but does not in itself lead to participation. So voice has multiple 'meanings' and can be interpreted in different ways such as being seen as a countervailing source of power on management action or perhaps part of a mutual gains process (Dundon *et al.*, 2004). The current business narrative is that organisations need to take the high road with high value-added operations or be dragged down into competing for low value-added jobs that are in danger of moving abroad (Handel and Levine, 2004). As Strauss (2006, 778) observes, getting workers voice 'provides a win-win solution to a central organisational problem – how to satisfy workers' needs while simultaneously achieving organisational objectives'. However, theory and practice can diverge (Wilkinson and Fay 2011).

Voice potentially allows staff to influence events at work, and can bring together collective and individual techniques into one framework. HR managers typically play an important role in the choices made about employee voice, in identifying the options available, forming alliances with line managers and devising strategies for implementation. Generally a broad mix of factors shape management choice. For some, rational choice was about satisfying employee expectations, particularly when faced with tight labour markets; others equate choice with their own understanding of corporate and organisational objectives. Some managers may feel that they have no choice, either because employees demanded 'a say' or due to market pressures and new legislative requirements (Dundon *et al.*, 2004). On the whole, however, management decides whether or not workers have a voice, and it is managers rather than employees who decide what mechanisms to utilise. Thus the main aim of this approach to voice reflects a management agenda concerned with increasing understanding and commitment from employees and securing an enhanced contribution to the organisation (Marchington and Wilkinson, 2012).

Employee engagement

The Chartered Institute of Personnel and Development, the professional body for UK HR specialists, defines engagement as 'a combination of commitment to the organisation and its values, plus a willingness to help out colleagues. Engagement is something the employee has to offer: it cannot be 'required' as part of the employment contract.' However, a number of reports suggest there is value in engaging staff more. For example, a Watson Wyatt study (2009) indicated that a company with highly engaged employees achieves a financial performance which might be as much as four times better than those with poor engagement. In the UK the Macleod Report entitled *Engaging for Success* was designed to open a national discussion on the subject but the assumption behind this was not to debate the merits of the idea, but to work out how best to implement it. While one-size-may-not-fit-all, Purcell (2010) points out that there are key common factors linking the experience of work to organisational commitment and engagement for most employees. Drawing from the Workplace Employee Relations Survey data he notes that it is possible to identify eight main occupational groups and find the factors significantly linked to positive commitment for all or most occupations. These are:

- employee trust in management (significant for all occupations)
- satisfaction with the work itself (significant for 7 out of 8 occupations)
- satisfaction with involvement in decision-making at the workplace (significant for 6/8 occupations)
- quality of relationships between management and employees, sometimes called 'employee relations climate' (significant for 5/8 occupations)
- satisfaction with the amount of pay received (significant for 5/8 occupations)
- job challenge (significant for 5/8 occupations)
- satisfaction with sense of achievement from work (significant for 4/8 occupational groups).

These factors can be seen as the basic building blocks for employee engagement. From the WERS data the number who are 'fully engaged', defined as scoring highly on every dimension, is often less than one-in-five. The expectation would be that the bulk of employees in a well-functioning firm would be 'engaged' (i.e. score 4) and the median score would be over 3. Lower levels of engagement are more likely to be found where there is perceived unfairness in rewards, where there is bullying and harassment, and where people believe they are stuck in their jobs and feel isolated from open communications.

Conclusion

We began this chapter by considering what exactly is meant by employment relations as both a field of study and as a component of HR policy and practice, and outlined the various levels at which issues can be studied (macro, micro and meso). Specifically, we emphasised the value of exploring employment relations patterns at the meso level, as it allows sensitivity to the embeddedness of firms within a particular regulatory and institutional environment which influences the dominant approaches and norms that characterise a particular sector. This context guided our exploration of several broad employment relations themes. Firstly, the complex and distinctive nature of the employment relationship and its various dimensions were reviewed. It was suggested that the employment relationship has economic, legal, political and psychological aspects, and that it is also indeterminate, unequal and dynamic. However, the nature of the employment relationship in a construction context seems even more complex because of the emphasis upon maintaining 'flexible firms', the project-based nature of the industry, and the transient and fluid nature of the workforce. Employment arrangements considered 'atypical' or 'non-standard' in many other sectors, such as self-employment and fixed-term working, are often commonplace in construction. This was followed by a discussion around the theories of 'frames of reference' and how these can be used to describe various perspectives regarding the nature of the employment relationship, and in particular the differences between unitarist and pluralist views. Again, the diversity of employment relationships in the construction sector creates an additional level of complexity in managing employment relations. While at enterprise level, firms may seek to minimise direct labour and employment responsibilities associated with direct employment, the challenges of managing convergent and divergent interests do not disappear: arguably they are even more important especially at a sectoral level. We then explored some of the key actors and parties involved with the conduct and regulation of employment relations, namely management, the state, and trade unions and outlined some of the main choices available to each in this regard. The preponderance of small firms, blurred organisational boundaries, and multiplicity of employment arrangements all raise distinctive challenges for employers, trade unions and the state in order to develop and sustain positive employment relations. We concluded by outlining several contemporary employment relations issues, and illustrated the complexity and potential challenges associated with the effective management of the employment relationship in construction.

References

Ackers, P., Smith, C. and Smith, P. (eds) (1996) *The New Workplace and Trade Unionism, Routledge*, London.

Ackers, P. and Wilkinson, A. (2003) *Understanding Work and Employment: industrial relations in transition*, Oxford, OUP, Chapter 1.

Atkinson, J. (1984) Manpower strategies for the flexible organisation, *Personnel Management*, 28–31.

Atkinson, J. and Meager, N. (1986) *Changing Working Patterns: How Companies Achieve Flexibility to Meet their Needs*, National Economic Development Office.

AusStats (2011) Employee earnings, benefits and trade union membership, Available at http://www.ausstats.abs.gov.au/Ausstats/subscriber.nsf/0/2C82C7433748647 2CA25788700131F58/$File/63100_aug%202010.pdf.

BBC News (2006) Strike halts work at Dubai tower, 23 March, Available at http://news.bbc.co.uk/1/hi/business/4836632.stm.

BBC News (2009) Q and A: What is the dispute all about? 4 February, Available at http://news.bbc.co.uk/1/hi/business/7860622.stm.

Beardwell, I. and Holden, L. (2001) Human *Resource Management: a contemporary approach*, 3rd edn, FT Prentice Hall.

Beardsworth, A., Keil, E., Bresnen, M. and Bryman, A. (1988) Management, transience and subcontracting, Journal of Management Studies, 25(6), 603–25.

Beaumont, P.B. (1995) *The Future of Employment Relations*, London, Sage.

BIS (2011) Trade Union Membership 2010, Department of Business innovation and Skills, Available at http://stats.bis.gov.uk/UKSA/tu/ TUM2010.pdf.

BLS (2011) Union Members Summary, Bureau of Labour Statistics http://www.bls. gov/news.release/union2.nr0.htm.

Blyton, P. [point] and Turnbull, P. (2004) The *Dynamics of Industrial Relations*, Macmillan.

Bray, M., & Waring, P. (2009) The (continuing) importance of industry studies in industrial relations, [comma] *Journal of Industrial Relations*, 51(5), 617

Budd, J. (2004) *Employment with a Human Face,* Cornell University Press.

Clegg, H. (1975) Pluralism in industrial relations, *British Journal of Industrial Relations*, 13(2).

Clegg, H. (1979) *The Changing System of Industrial Relations in Great Britain*, Oxford, Blackwell.

Clarke, L. and Hermann, G. (2007) Skills shortages, recruitment and retention in the housebuilding sector, *Personnel Review*, 36(4), 509–27.

Coffeey, M. and Langford, D (1998) The propensity for employee participation by electrical and mechanical trades in the construction industry, *Construction Management and Economics*, 16(5), 5434–552.

Colling, T. and Terry, M. (2010) *Industrial Relations: Theory and Practice*, Wiley.

Croucher, C. and Druker, J. (2001) Decision taking on human resource issues, *Employee Relations*, 32(1), 55–74.

Dainty, A. (2007) People and Culture in Construction: A Reader, Abingdon, Taylor and Francis.

Deakin, S. and Koukiadaki, A. Governance processes, labour management partnership and employee voice in the construction of Heathrow terminal 5, Industrial Law Journal 38 (2009): 365–89.

DETR (1998) Rethinking Construction and Accelerating Change, London.

Doloi, H. (2009) Analysis of pre-qualification criteria in contractor selection and their impacts on project success, Construction Management and Economics, 27(12), 1245–63.

Dundon, T. (2002) Employer opposition and union avoidance in the UK, *Industrial Relations Journal*, 33(3), 234–45.

Dundon, T., Wilkinson, A., Marchington, M. and Ackers, P. (2004), The meanings and purpose of employee voice, *International Journal of Human Resource Management,* Vol.[point] 15 (6), 1150–71.

Druker, J. and Croucher, R. (2000) National collective bargaining and employment flexibility in the Euopean building and civil engineering industries, *Construction Management and Economics*, Vol. 18 No. 4, 699–709.

Druker, J. and White, G. (1996) Managing *People in Construction*, IPD, London.

Druker, J., White, G., Hegewisch, A. & Mayne, L. (1996) Between hard and soft HRM: human resource management in the construction industry *Construction Management and Economics*, *14*(5), 405–16.

Edwards, P. (2003) The employment relationship in the field of industrial relations, in *Industrial Relations: Theory and Practice*, Blackwell.

Ewing, K. (2009) *Ruined Lives: blacklisting in the UK construction industry – a report for UCATT*, Institute of Employment Rights, London.

Ferner, A. and Hyman, R. (1998) *Industrial Relations in the New Europe*, Oxford: Blackwell.

Flanders, A. (1965) *Industrial relations – what is wrong with the system?* Faber and Faber.

Forde, C., MacKenzie, R. and Robinson, A. (2009) Built on Shifting Sands: Changes in Employers' Use of Contingent Labour in the UK Construction Sector, *Journal of Industrial Relations*, 51, 5, 653–67.

Fox, A. (1966) *Industrial Sociology and Industrial Relations*, Royal Commission Research Paper Number 3, London, HMSO.

Fox, A. (1974) *Beyond Contract: Work, Power and Trust Relations*, Faber and Faber.

Gall, G. 2009 Union Revitalisation in Advance Economies, Palgrave Macmillan.

Green, S. D. (2002) The human resource management implications of lean construction: critical perspectives and conceptual chasms, *Journal of Construction Research*, 3(1), 147–66.

Green, S. (1999) The missing arguments of lean construction, *Construction management and economics*, 17, 133–7.

Green, S. and May, S. (2003) Re-engineering construction: going against the grain, Building Research and Information, 31(2), 97–106.

Guest, D. (1995) Human resource management, industrial relations and the trade unions in Storey, J., Human Resource Management: A Critical Text, Routledge.

Guest, D. and Hoque, K. (1994) The good, the bad and the ugly, *Human Resource Management Journal* 5(1), 1–14.

Guest, D. and Conway, N. (1999) Peering into the black hole: the downside of the new employment relations in the UK, *British Journal of Industrial Relations*, 37(3), 367–89.

Hale, D. (2008) Labour disputes in 2007, Office for National Statistics.

Handel, M. J., & Levine, D. I. (2004) Editors' Introduction: The Effects of New Work Practices on Workers, *Industrial Relations*, 43(1), 1–43.

Heery, E. (2002) Partnership versus organising: alternative futures for British trade unionism, *Industrial Relations Journal*, 33(1), 20–35.

Heery, E., Bacon, N., Blyton, P. and Fiortio, J. (2008) Introduction: the field of industrial relations in Blyton, P., Bacon, N., Fiortio, J. and Heery, E., Sage Handbook of Industrial Relations, London, Sage, 1–32.

Hillebrand, P. and Cannon, J. (1990) The Modern Construction Firm, Macmillan, London.

Hyman, R. (1975) Industrial Relations: A Marxist Introduction, Macmillan, London.

Hyman, R. (2004) Whose (social) partnership in Stuart, M. and Martinez-Lucio, M. (2005) Partnership and Modernisation in Employment Relations, Routledge.

ILO (2001) The Construction Industry in the Twenty-First Century: Its Image, Employment Prospects and Skill Requirements, International Labour Office, Geneva.

Katz, H. and Colvin, A. (2011) Employment relations in the United States in Bamber, G. Lansbury, R. and Wailes, N (eds.) International and Comparative Employment Relations, Sage, London.

Kochan, T. and Osterman, P. (1994) *The Mutual Gains Enterprise*, Harvard Business School, MA.

Leung, J., James, K., Mustata, R. and Bonaci, G. (2010) Trade union strategy in Sydney's construction union, International Journal of Social Economics, 37(7), 488–511.

Lewin, D. (2010) Mutual gains in Wilkinson *et al.*, Handbook of Organisational Participation.

MacKenzie, R., Forde, C., Robinson, A., Cook, H., Eriksson, B., Larsson, P. and Bergman, A. (2010) Contingent work in the UK and Sweden: evidence from the construction industry, *Industrial Relations Journal*, 41(6) 603–21.

MacKenzie, S., Kilpatrick, A. R. and Akintoye, A. (2000) UK Construction Skills Shortage Response Strategies and an Analysis of Industry Perceptions, *Construction Management and Economics* 18(7): 853–62.

Marchington, M. P., Cooke, F., Hebson, G. (2008) Human resource management across organizational boundaries, in Wilkinson, A., Redman, T., Snell, S. and Bacon, N. (eds), Sage Handbook of Human Resource Management, London: Sage.

Marchington, M. and Wilkinson, A. (2012) Human *Resource Management at Work*, CIPD, London.

McGrath-Champ S. and Rosewarne S. (2009) Organisational change in Australian building and construction: rethinking a unilinear 'leaning' discourse, Construction Management and Economics, Vol. 27:11, 1111–28.

Nolan, P. and Wood, S. (2003) Mapping the future of work, *British Journal of Industrial Relations*, 41(2), 165–74.

Purcell, J., Sisson, K. (1983) Strategies and practice in the management of industrial relations, in Bain, G. (eds), *Industrial Relations in Britain*, Blackwell, Oxford.

Purcell, J. (2010) Building employee engagement, *ACAS Policy Discussion Papers*, (January).

Raidén, A., Pye, M. and Cullinane, J. (2007) The nature of the employment relationship in the UK construction industry in Dainty, A. *et al.* (eds), People and Culture in Construction, Spon.

Rubenstein, S. A. and Kochan, T. (2001) Learning from Saturn, *Relations Industrielles/Industrial Relations*, 56(5).

Rubery, J., Carroll, M., Cooke, F., Grugulis, I. and Earnshaw, J. (2004). Human resource management and the permeable organisation: The case of the multiclient call centre, Journal of Management Studies, 41(7), 1199–222.

Simms, M. and Charlwood, A. (2010) Trade Unions: Power and Influence in a Changing Context in Industrial Relations: Theory and Practice, Wiley.

Sisson, K. (1993) In search of human resource management, *British Journal of Industrial Relations*, 31(2) 201–10.

Sisson, K. (2010) *Employment relations matters,* Available at www2.warwick.ac.uk/fac/soc/wbs/research/irru.

Strauss, G. (2006) Worker Participation: Some Under-Considered Issues, *Industrial Relations* 45:4, 778–803. October.

Stuart, M. and Martinez-Lucio, M. (2005) *Partnership and Modernisation in Employment Relations,* Routledge, London.

Uppal, S. (2010) Unionization 2010, Perspectives – Statistics Canada, www.statcan.gc.ca/pub/75-001-x/2010110/pdf/11358-eng.pdf

Vaid, K. N. (1999), Contract Labour in the Construction Industry in India, in D.P.A. Naidu (Ed.). *Contract Labour in South Asia,* Geneva: Bureau for Workers' Activities, ILO.

Waddington, J. (1992) Trade union membership in Britain 1980–1987, *British Journal of Industrial Relations,* 30(2), 287–342.

Webb, S. and Webb, B (1920) *The History of Trade Unionism 1866–1920,* London, Longman.

Wilkinson, A. and Fay, C. (2011), New times for employee voice?, *Human Resource Management,* 50(1), 65–74.

Wilton, N. (2010) *An Introduction to Human Resource Management,* Sage.

Winch, G. (1998) The growth of self-employment in British construction, *Construction management and economics,* 16(5), 531–42.

9 Illusions of equity, procedural justice and consistency: a critique of people resourcing 'best practice' in construction organisations

Ani Raidén and Anne Sempik

Introduction

This chapter critically examines the notion of HRM 'best practice' as applied to people resourcing in the construction industry. In smaller organisations management practices are informal and relatively simple. Thus, the discussions in this chapter focus on issues more relevant to larger and more complex construction companies and their professional employees.

Alongside learning and development and employment relations, people resourcing activities tend to focus on operational human resource management within organisations. The aim of people resourcing is described by many authors (including Raidén *et al.*, 2009: 51) as intended to ensure that an organisation has:

> the right number of people with the right skills, behaviours and attitudes in the right place at the right time.

The management activities designed to achieve this include, for example, recruitment and selection, performance management and team deployment. These activities are generally presented as largely straightforward, uncontested areas of the employment relationship and it is assumed that a best practice approach can be applied to them. Best practice approaches are usually characterised by a concern for equitable treatment of every employee and considerations of procedural justice (Rawls, 1999). The implicit message underpinning best practice approaches is that they will result in fair and positive outcomes for both employer and employees, hence the association with high commitment management.

HRM best practice forms a framework to structure our discussion of the key people resourcing activities: selective hiring and sophisticated selection, performance review and appraisal, teamworking, and employment security and internal labour markets. Towards the end of the chapter we present a case study of people resourcing within a large UK-based contractor that helps illuminate the challenges of best practice HRM in resourcing.

Our intention is to stimulate an interest in the challenges of resourcing within construction organisations. We argue that while the best practice approach can be useful in alleviating limitations in subjective decision-making, people involved in resourcing activities (i.e. managers and employees) bring with them complex variations in perception and motives.

HRM best practice

There are two ways of approaching HRM: a 'best practice' and 'best fit'. 'HRM Best Practice' is associated with 'high-commitment' human resource management (Huselid, 1995; Pfeffer, 1998). In many cases it is seen as a prescription for a more successful way of managing people and organisations. Some writers have associated what we are talking about with improved organisational performance and so they are also often referred to as 'high performance work systems' (Appelbaum *et al.*, 2000). The HR practices included under the banner of best practice are: employment security and internal labour markets; selective hiring and sophisticated selection; extensive training, learning and development; employee involvement and participation; self-managed teams/teamworking; performance review, appraisal and career development; high compensation contingent on performance; reduction of status differentials and work–life balance (Marchington & Wilkinson, 2008: 94) (for a comprehensive discussion of HRM best practice please consult this literature). Best practice advocates that all of these should be applied to all organisations. 'Best fit' by contrast suggests that organisational practice should be adjusted to organisational circumstances and environmental situations (Purcell, 1999: 33).

A desired outcome of HRM best practice is employee engagement. This is a topic area that has achieved a level of interest previously accorded only to talent management (and before it the concept of HRM itself). In part, this has been led by the professional body for HRM in the UK: the Chartered Institute for Personnel and Development (CIPD), who first sponsored research into what Purcell *et al.* (2003) described as the 'black box.' Descriptions of the concept vary with some suggesting it is an attitude where others see it as a set of behaviours (Kahn, 1990; Robinson *et al.*, 2004; Saks, 2006). Despite this ambiguity, employee engagement is a topic that enthuses industry and research organisations (for example Gallup, 2010) offering an explanation of why some organisations perform better than others.

The importance of employee engagement to resourcing is that research evidence suggests engaged employees work harder and are willing to invest more discretionary effort in their work and this can be measured in improved organisational performance. Furthermore, it also suggests that employers can create an environment in which employees will want to work harder and thus considerations of how this can be achieved underpin the resourcing effort.

Measures of better 'HR health', for example in lower than average absence rates among professional employees in the UK construction industry (see CIPD, 2009), support a contention that these employees are engaged. However, one might question the direction of their engagement; is it to their employing organisation, their profession or to the eventual realisation and completion of a project with which they are associated? (See literature on knowledge workers and their loyalties for example in Scarbrough,1999; Scarbrough and Carter, 2000.)

In this chapter we shall critically analyse those elements of HRM best practice that are directly relevant to people resourcing:

- Selective hiring and sophisticated selection
- Performance review and appraisal
- Teamworking
- Employment security and internal labour markets.

The other elements of best practice (such as training, learning and development) are addressed elsewhere in this book.

Selective hiring and sophisticated selection

The aims of recruitment and selection are: attracting the right candidates; choosing the most suitable from amongst them; making an offer of employment and; effectively on-boarding or inducting them into the organisation with the intention of retaining them in the medium to long-term. Unfortunately often recruitment and selection processes are too focused on attraction and selection, and hence only partially achieve these aims (in terms of hiring new employees; but not retaining them). 'Induction crisis', poor performance, poor attendance, high labour turnover, high levels of grievance and discipline issues are frequently *symptoms* of problems in recruitment and selection. Here we explore, first 'recruitment and selection' as a whole, before reviewing 'recruitment' and then 'selection' separately.

Typically, recruitment and selection activities are presented as a systematic process in which the aim is objective decision making at every stage. Competency frameworks are a contemporary development to supplement or replace traditional person specifications and are said to make the process more objective (Bloisi, 2007). Competency frameworks allow an organisation to explicitly communicate to its employees and potential future employees the kinds of behaviours and performance that will be valued and recognised (i.e. those that help the organisation achieve its strategic goals). However, despite the espoused benefits in terms of objectivity, the use of competency frameworks has been met with accusations of 'cloning' (see for example Sparrow and Bognanno, 1993) and concerns over equal opportunities/ diversity. Where a competency framework is developed in an overwhelmingly male environment (such as construction) many of the preferred behaviours are likely to be masculine in style and women may find it difficult to emulate this behaviour. Thus, use of the competency framework may indirectly discriminate against a particular population. Moreover, where women exhibit some masculine competencies (such as assertiveness), interpretation of their behaviour might be skewed in the light of their gender (Biernat and Fuegen, 2001): an assertive woman can be perceived by men and other women as being aggressive.

Competencies also tend to focus on inputs (what people do and how they do things) rather than outputs (what they achieve) whilst the reverse is true for much of modern HRM. However, researchers such as Cheng *et al.*(2005; see also Dainty *et al.*, 2005) within the construction management community actively incorporate both input (underlying behaviours) and output-based criteria (performance/technical) in their search for a competency profile of a superior manager. Nevertheless, application of even this kind of balanced approach is not without practical difficulties in the development and use of any competency framework. The assessment and interpretation of behaviours is still mostly subjective and requires ongoing review and evaluation, which are likely to be time-consuming. Also, where a project is HR led, gaining line management buy in may be difficult (see CIPD, 2003, where all of these issues are explored extensively).

Recruitment

Recruitment is generally regarded as the activities which culminate in people applying for a job. This typically involves: job analysis; preparation of job descriptions and person specifications which describe both the job and the essential and desirable knowledge, skills, experience and possibly behavioural characteristics of the ideal job holder; and then advertising the job in some way, often in paper media or increasingly online. Alternative methods of recruitment include the use of employment agencies, recruitment consultants, headhunters and, less formally, word of mouth. Recruitment consultants and headhunters are often used in a tight employment market where specific skills are sought or for more senior positions where there are perhaps only a few people who have the necessary skills and experience to fulfil a particular role. Word of mouth and 'direct approaches to potential employees by the eventual employer' are also a response to a perceived tight labour market. To be successful these more personal methods rely on good knowledge of the personal characteristics of those approached. The construction industry routinely makes extensive use of informal methods of recruitment as many managers responsible for hiring have knowledge of who could be approached through personal networks (Druker and White, 1996; Lockyer and Scholarios, 2007; Raidén *et al.*, 2009: 132). The drawback of informal methods is that they do not stand up well to scrutiny from an equal opportunities perspective. Research in the UK suggests that friendships and social networks are usually comprised of one ethnic group (Phillips, 2005).

The implication of this is that the pool of candidates recruited for any position by these methods will only be drawn from as wide a part of society as the professional network of the managers responsible for the recruitment campaign.

Selection

Selection is generally regarded as the activities through which the pool of candidates is reduced until job offers can be made to those who most closely match the criteria established in the person specification (see for example Bloisi, 2007; Taylor, 2010; Chamorro-Premuzic and Furnham, 2010). Traditionally the three main approaches to selection make up 'the classic trio'. This refers to the common practice of requiring applicants to (i) complete an application form or send in a CV, (ii) attend one or two interviews and (iii) have references taken up (Cook, 2004: 3).

Iles (1999) refers to the systematic process as the psychometric perspective. This is based on establishing the validity and reliability of selection methods, i.e. the extent to which they predict consistently the likelihood that an individual will perform the job effectively. Selection methods are then applied as tools in a neutral process that does not recognise the social and political contexts (Iles, 1999; Bloisi, 2007; Searle, 2003) and assumes that evidence gathered as part of the selection process can be evaluated dispassionately and objectively. Others make reference to 'the social process perspective' (Newell, 2006; Newell and Rice, 1999), which sets recruitment and selection in its social context and takes account of complex range of factors which may influence decision-making. These include but are not limited to:

- perceptual bias in selectors
- the bias and personal motives that influence the analysis of jobs; and the content of job descriptions, person specifications and recruitment advertisements
- the motivations of applicants.

Thus the social process perspective enables us to see that recruitment and selection are just as likely to be subject to the 'garbage can' theory of decision making 'in which a multiplicity of issues and differences of power and perception are thrown together and help shape outcomes' (Watson, 2002: 321).

Newell (2006) argues that, because the systematic process asserts a best way of doing a job, it usually ends up being offered to someone similar to the person who did it before, perpetuating existing gender and ethnic segmentation. It thereby presents a facade of fairness hiding the reality of subjective bias in the process.

One aspect of resourcing practice that seeks to eradicate the potentially negative effects of subjective decision-making is job analysis.

Job analysis

Job analysis refers to the process by which the component parts of any job are identified and information is gathered to enable critical evaluation. From a HRM point of view, job analysis is 'one of the building blocks of an organisation, providing a blueprint of how the organisation works' (Searle, 2003: 24) and thus an activity of strategic importance.

Data collection can be carried out by observing or filming the job holder during a representative period of activity, by completion of a diary or questionnaire, through interviews or focus group discussions with job holders, by analysis of critical incidents or through repertory grid analysis (Searle, 2003; Cook, 2004). Interviews with job holders allow the job analyst to understand

> plans, intentions and meanings. The person who sees his or her work as 'laying bricks' differs from the person who sees it as 'building cathedrals' (Cook, 2004: 23).

There is an increasing focus within organisations to establish not only the task elements of a job but also *how* a job is done (Searle, 2003). The purpose of job analysis is to create a clear understanding of the role tasks and behaviours so that it is possible to write a job description and a person specification which will allow recruitment advertising and selection criteria to be developed. Managerial interpretation of information may influence this process. Some managers will prefer a particular style of work in their subordinates (Dainty and Bagilhole, 2006: 104) and thus, if asked to select a subordinate to be interviewed, keep a diary, or be filmed doing the job, they will select those who best fit their own preferred way of doing things thus introducing personal bias into the analysis. Their style may or may not suit the organisation's current context.

The systematic process approach to recruitment and selection advises that job analysis should precede every recruitment effort (Martin *et al.*, 2010: 123). The reasons for this are first to ensure that the job is still relevant and required and that the work cannot be shared out amongst existing employees. Second, it is important to review the job requirements at regular intervals from an equal opportunities perspective. There is often a temptation when a job is to be filled to replace 'like with like' and so if a job has been carried out by one person on a full time basis consideration is rarely given as to whether two people could do it through job share for example. Since the majority of jobs in construction are occupied on a full-time basis by able-bodied men from the ethnic majority, this 'like-for-like' attitude tends to perpetuate existing labour market segregation and stands to limit the labour supply from which the organisation can draw (Newell, 2006). Frequently however, there is insufficient time to re-analyse a job due to pressure to fill a vacancy and the opportunity to review practice and outcomes is missed.

There is little, if any, research of organisational practice in construction in this area.

Performance review and appraisal

Performance appraisal is a tool for determining and setting performance expectations, and reviewing and appraising performance (Latham *et al.*, 2007). More specifically, within the HR profession (after CIPD, 2010a) appraisal is commonly viewed as:

> an opportunity for the individual and those concerned with their performance, most usually their line manager, to get together to engage in a dialogue about the individual's performance, development and the support required from the manager. It should not be a top down process or an opportunity for one person to ask questions and the other to reply. It should be a free flowing conversation in which a range of views are exchanged.

This view is predicated upon line managers knowing the work of their employees and having the skills and interest to conduct effective appraisals, and that the employee whose performance and development is under review understands and engages fully in the process. These aspects cannot be taken for granted (Longenecker, 1997). Hence, it is not surprising that performance appraisal is a contested area of people resourcing. Since the appraisal is often associated with career advancement and/ or pay awards, employees may not feel it is in their best interests to be open and honest about aspects of their job with which they have struggled as they may feel that this would affect possible opportunities for promotion or pay award. Line managers may also feel defensive about the process, especially if they anticipate a difficult meeting. Several decades ago, McGregor (1957: 90) suggested that the reluctance to carry out appraisals was due to discomfort at 'playing God': judging the worth of a subordinate and having to communicate the judgement, whilst at the same time being responsible for that subordinate's development. The problems go further than this however. For example, there may be disagreement between a line manager and an employee as to what targets were agreed and how they would be assessed, or there may be disagreement about whether a target has in fact been achieved (Longenecker, 1997). The CIPD (2010b) has rather given up on this aspect and merely report that good line managers are invariably competent appraisers of performance.

One aspect of performance appraisal that has received more positive attention is the 360° appraisal. This is when an individual receives feedback on their performance from a number of sources: their line manager; peers; subordinates; customers and; themselves. Supporters of this approach say that it improves upon line manager feedback as it is less subject to the vagaries of the personal relationship between two people and allows a more rounded picture of the individual to emerge. Peers' opinions may be more accurate than that of the line manager (Redman, 2006: 161). The information can also be presented neutrally to the recipient, perhaps after they have self-rated the same set of questions. Discussion can be then focused on the discrepancies between the self- and other assessments, and on any differences in perception between groups of assessors. This can contribute to a sense of ownership and autonomy amongst employees (Redman, 2006), which in turn may lead to greater commitment and engagement with organisational goals.

For an industry such as construction where employees are working in locations with little contact with their actual workplaces, 360° appraisal 'can provide a more meaningful appraisal' (Redman, 2006: 162) taking into account client and other stakeholder feedback. Most 'end of project review' mechanisms collect customer feedback on project team performance, which could be fed into the team's appraisal and then filtered down to individual level. Moreover, such existing data collection instruments could be easily revised to include more specific questions on what is considered important in an organisation, such as key behavioural competencies.

One significant challenge in the 360° system is that it is rather costly to operate (and hence we do not know of many construction organisations engaging with this form of appraisal). Developments in human resource information technologies, such as Web 2.0 that allows two-way communication and feedback (Martin *et al.*, 2009), may offer opportunities to extend the practice of 360° appraisal.

Teamworking

Teamworking is an area of people resourcing practice that is central to project-based organisations. It is an inherent part of construction operations, but also increasingly popular way of organising work within more stable environments as project-/ teamwork becomes integral to strategy, for example, in research and development, quality assurance and managing change. Teamworking (sharing knowledge) is said to hold the key to success in knowledge-based organisations and knowledge-based roles, such as general/ line managers and HR professionals (Taylor, 2010: 457). Despite this, team deployment is usually excluded from considerations of people resourcing issues and activities, and rarely features in models/ frameworks for good practice. People resourcing generalists like Taylor (2010) appear to regard it as more of an operational issue, and therefore line management concern, rather than a HRM activity or a topic attached to altering work arrangements in order to improve efficiency and quality. Although the most recent edition of the CIPD text (Taylor, 2010) highlights workforce mobilisation and teams as part of the major and fundamental objectives of resourcing and talent planning, team deployment is not discussed in any detail. Here, construction organisations have an opportunity to lead the field in development and showcasing of innovative approaches to team deployment. However, we fear this will be an opportunity missed as, at present, much of the emphasis in analysis and recommendations for best way forward in team deployment/ project staff allocation tends to be on automation. This is driven by a desire to establish an *'optimal human resource management allocation'* (Chen *et al.*, 2008: 804) in order to reduce cost.

A more sensible approach is perhaps to seek well-informed decisions – research has confirmed that employees (and managers) will accept and be satisfied with less desirable solutions when they understand the reasons for their manager's actions (i.e. transparent communication is in place) and when they trust that their needs and preferences will be taken into account (Raidén *et al.*, 2009: 141). Raidén *et al.*(2006, 2009) seek a balanced approach to assembling construction project teams, advocating the value of qualitative (subjective) forms of decision-making and knowledge (as also discussed within human resource planning later in this chapter) but in combination with systematic, organisation-wide, IT-enabled technologies. Such an approach highlights horizontal integration of the kinds of people resourcing activities discussed here (e.g. human resource planning, performance management and team deployment) with training and development activities as well as employment relations issues such as employee involvement. Importantly, dismantling (and redeployment) of teams is an explicit element of this model so that crucial project learning is not lost and a strategic view of the deployment process is maintained.

Employment security and internal labour markets

The conditions for employment security and the development of the internal labour market are created by strategic consideration of flexible employment strategies, management of change and human resource planning together with operational people resourcing practices such as 'exit management'.

Flexible employment strategies

Flexible employment strategies were first defined by Atkinson (1984) in a model that could have been based on the resourcing practice of the construction industry. The model proposes a three-tier form for use of labour. At the centre is a core group of employees, the most qualified or skilled and therefore valuable talent, who benefit from permanent employment contracts and the benefits that accompany this (e.g. access to training and development, contingent reward, pension provision). Immediately outside the core is a peripheral workforce; this group may have contracts of employment with the organisation but often in a variety of part-time or temporary roles. The peripheral labour supply can be increased or decreased easily and its presence provides a buffer for the core.

Further still from the core of the organisation is a second peripheral group of workers who may be supplied by agencies or sub-contractors, and include the self-employed and people who may be associated with the organisation by dint of providing services which have been outsourced (typical functions to have been outsourced are payroll and IT). These latter types of flexible employment strategies are referred to as distancing strategies.

The periphery typically provides most flexibility as it is seen as either not as valuable in terms of skills as the core, possibly is more easily available in the employment market or, in the case of distanced workers, consists of someone else's employees and can be 'let go' in times of economic downturn or between projects and rehired when the economy picks up or the next project presents itself.

Atkinson's model has stimulated debate about whether this form of employment actually describe trends or seeks to predict employment trends of the future, and the extent to which organisations adopt it as a strategic tool or a reactive response to changes in the external environment (see Pollert, 1991; Legge, 1995). Research reported by Torrington *et al.*(2002) suggests that, in the UK, numerical flexibility is the most common form and is adopted as a short-term measure. Where they found evidence of functional flexibility it seemed to be part of a more strategic approach.

The key arguments in favour of flexible employment strategies are that they allow productivity gains through the more efficient use of labour. Whilst these are compelling, an alternative point of view suggests that such practices may create long-term damage to an organisation's competitive position through loss of expertise and impaired motivation (Torrington *et al.*, 2002). There are also widely recognised potential human costs in terms of lost entitlements, poor training, abuse by unscrupulous employers and employment uncertainty. The many problems other associated with the 'hollowed-out' firm are discussed in other chapters in this book.

Within the construction industry's professional and managerial groups (such as quantity surveyors and project managers) distancing strategies are widely adopted, each group is aware that whilst it may be a peripheral workforce to the organisation which provides their contract of employment it is part of the core workforce as far as the organisation to which work is contracted is concerned. Thus, loss of expertise and erosion in motivation are less likely issues for this group. Indeed, many individuals enjoy the flexibility (and status) provided to them too by this model.

Managing change

Managing change is about the management of business and HRM activities so that organisation progressively evolves. Change management is no longer contained within planned programmes or discrete episodes during which change happens. It is an inherent and ongoing feature of the contemporary business environment; and construction is arguably more prone to constant change *within* its operating environment as well as influenced by the bigger economic, societal, legislative, technological, political and other forces of change than many other industries.

People resourcing is subject to change too. It is an increasingly 'watched over' function of HRM in organisations. Resourcing activities are increasingly subject to constantly developing legislation designed to protect employees/ prospective employees and agency staff, and extend to a wide range of equality measures. These changes are partly driven by demographic changes in the population and employment markets, and associated societal change in value systems. As noted earlier, they affect people resourcing practices, for example in how managers recruit and select staff, but increasingly legislation also affects the strategic considerations. Numerically flexible strategies were attractive to organisations when it was possible to dismiss temporary workers relatively easily; less so when they enjoy the same employment protection as permanent employees.

Increasing legislation may also drive the operational (compliance) approach many organisations take on resourcing and thus cause them to neglect the strategic role of the function. For example in recruitment and selection, managers view documenting evidence (of a candidates ability to respond to questions) as a 'protective measure' in case of a later claim. Hence the value in aiding/ supporting decision making is diminished, where really the value of job descriptions and person specifications is realised in introducing a level of objectivity into an inherently subjective process.

A key technology in supporting change in the people resourcing effort is the development of Human Resource Information Systems (HRISs). These provide an electronic database for the storage and retrieval of employee and organisational data that offers the potential for flexible and imaginative use of this data (Tansley *et al.*, 2001: 354). The ways and levels of practice within construction organisations vary greatly (Raidén *et al.*, 2001, 2008). While research in the field is limited; findings are united in that given the highly mobile and transient nature of the construction workforce, HRIS offer a more reliable, accurate and accessible means of human resource planning, reducing labour turnover and targeted training and development (Ng *et al.*, 2001; Raidén *et al.*, 2001, 2008). An organisation that employs temporary organisational structures and a mobile workforce, and requires rapid decision-making often on the basis of managerial judgement surely is ideally placed to benefit from a web-enabled system. However, there are challenges to enacting such an argument: research suggests HR professionals (and line managers) may not have the necessary skills and knowledge to analyse and interpret data and information (Lawler and Mohrman, 2003; Williams *et al.*, 2009) and the opportunity to draw down strategic information about human resource planning or team deployment is often held back by concerns regarding the integrity, reliability and consistency of data (CIPD 2004; CIPD 2005).

Human Resource planning

One of the practical ways of managing change and internal labour markets is human resource planning. Human resource planning, or manpower/ workforce planning is an area of debate within people resourcing and HRM/ strategic management more broadly. Most popular methods of planning have focused on collection and analysis of quantitative information and since it is difficult to come up with numerical plans that will remain accurate within fast changing business environments, managers and HR professionals have rightly viewed such plans with reservations. Despite the espoused benefits, it has never been a fashionable topic for research. However, new research suggests that workforce planning is becoming '*one of the hottest topics in town*' (Syedain, 2010: 24; Baron *et al.*, 2010) and there has been a shift in favour of qualitative methods, such as succession planning, flexible working and talent management (Baron *et al.*, 2010: 3).

The case *for* planning despite the fast changing nature of the business environment in construction relates to maintaining competitive advantage. Taylor (2010: 121–2) outlines two basic choices for employers: they can rely on managerial intuition, and react swiftly and decidedly as environmental change unfolds; or they can develop contingency plans for a variety of possible responses. The first approach closely correlates to the current model of management in construction: reactive 'fight for survival'. The latter, with its emphasis on future flexibility, enables organisations to position themselves ahead of the game for a variety of different circumstances.

Among those in favour of human resource planning, Bramham's (1988) model has been a popular framework. In this model, human resource planning is about assessing the external and internal environment in order to develop well-informed [contingency] plans to meet the business objectives. In short, Bramham's four-stage model focuses on:

1   Assessing and predicting future demand
2   Analysis and estimates on internal supply
3   Understanding and up-to-date information on external labour markets
4   Formulating responses to the above forecasts, analysis and other information in relation to organisation's business plans.

Broader coverage of the first three stages is provided in Taylor (2010) and in the context of construction specifically in Raidén *et al.*(2009). Usually, 'systematic techniques' (i.e. mathematical and statistical analysis) are preferred because of their apparent transparency and objectivity, and the fact that infinite numbers of variables can be introduced (limited only by the availability of data). For example, productivity trend analysis (see Abdel-Wahab *et al.*, 2008; Koskenvesa *et al.*, 2010) and quantitative manpower demand forecasting (Wong *et al.*, 2011) have received some attention in construction management research. However, another method that has been found particularly relevant for construction organisations (see for example the organisation case study later in this chapter), yet in the general literature is considered contentious, is managerial judgement. This approach to forecasting bases predictions and estimates on the subjective views of managers. The benefits of using managerial judgement centre on time efficiency and the requirement for little if any [additional] data collection; ability to incorporate intangible factors, such as local developments and employee preferences (Taylor, 2010: 108); and most importantly, use of expert knowledge. However, whilst there is no doubt that managers in any organisation are best placed to understand and react to what is happening in their environment, later in this chapter we discuss availability and use of information, perceptual bias and organisational politics which introduce significant potential for error. An approach which combines both qualitative and quantitative techniques and/or is conducted in a way that allows decision makers to debate the issues is therefore likely to result in more robust plans.

Traditionally the case against human resource planning has been founded on the argument that within the modern turbulent business environment it is unfeasible to forecast labour demand with any accuracy. While this view gained significant momentum in the nineties, more recently many organisations have realised the true benefits of planning ahead. Or perhaps more accurately, they found out the negative consequences of not planning their use of human resources (Watson and Wyatt, 2008a, 2008b). While the recession forced many organisations to cut costs in the short-term, and often release staff in order to achieve this, concerns about retaining and attracting critical skills and talent in the long term were brought to the fore. The previously rather polarised debate as to the value of human resource planning (i.e. whether to plan or not) now seems to engage more subtle discussion as to the types of techniques and methods most useful in the short- and long-term. In relation to 'exit', which we discuss next, value in planning relates to minimising compulsory redundancies.

Exit management

So far we have discussed different ways of getting people into an organisation and situating them in the most appropriate places within the company; exit refers to the release of personnel from organisations. Walsh and Lugsden (2009) provide a very useful overview of managing exit from organisations.

Their model includes different forms of employee-initiated exit, such as 'moving on' for change of career or retirement, and employer-initiated termination of employment, such as discipline and redundancy. They also address the more difficult forms of exit which may be the result of poor behaviour on part of the employer (constructive dismissal and failure to grant requests for flexible working or deferred retirement). Indeed, very often exit is seen as the negative end of the people resourcing cycle, a situation in which blame is assigned and people part company with ill-feeling.

Central to exit management is an understanding of the inherent conflict between organisational requirements and individuals' needs and preferences. Where carefully managed, taking into account how people feel, it may be possible to sustain a positive relationship beyond employment. For example, if an employee retires and feels supported through the transition in his/ her life, it is likely that they will enhance the organisation's reputation for a long time after by spreading positive messages.

Our focus here is on redundancy as many of the other forms of exit (dismissals in particular) are more about employment relations than people resourcing.

Redundancy

Strictly speaking, redundancy refers to 'a situation in which, for economic reasons, there is no longer a need for the job in question to be carried out in the place where it is currently carried out' (Taylor, 2010: 403). This specifies circumstances which usually involve an organisation with reduced workloads. Technological advances often initiate such change where parts of the production have been automated or outsourced, or administrative functions computerised. Also, organisational re-structuring may be used as the rationale for redundancies, although this is rarely a question of economics and more about managerial choice.

Curiously, although true redundancies build on economic rationale for the reduction in workforce, they are actually very costly (in the short-term). Redundancy payments themselves form only part of the expenditure. Successful business performance after downsizing requires HR practices which put the organisation back on its feet and these require investment. 'Survivor syndrome', where remaining employees feel guilty because their co-workers were chosen for redundancy and not them and/ or sense of loss over the departure of their colleagues, can have a significant negative impact on performance. Survivor syndrome may also mean increased levels of job insecurity.

Since (compulsory) redundancies commonly prove a traumatic experience for all parties involved: those being made redundant, the survivors, their managers and HR, they should be handled very carefully. In terms of good management practice, the key objective in considerations of redundancy should be how to *avoid* compulsory redundancies. In table 9.1 below we

Table 9.1: Alternatives to compulsory redundancies

	Short- to medium-term	Long-term
Avoiding redundancies	Wage reductions Part-time work Career breaks Redeployment Reduction of costs related to pay (e.g. cutting overtime and bonuses) Reduction of non-pay costs (e.g. office area/ space or hiring less expensive premises)	Workforce planning (being aware of changing business plans – reduce recruitment of 'at risk' groups) Flexible employment strategies (use of numerically flexible strategies, e.g. agency workers and fixed term contracts, and/ or multiskilling employees to facilitate redeployment)
When the workforce has to be reduced	Recruitment freeze	Call for volunteers Offer early retirement

highlight some mechanisms for managing the cost of the workforce in the short- to medium- and long-term.

However, situations where compulsory redundancies are the only option available do arise and here attention needs to be on (i) fair selection for redundancy and (ii) support for both those being made redundant and those who stay (the survivors) (see Bradley, 2010). Selection criteria often deemed to be fair and objective include length of service, qualifications/experience, disciplinary records, attendance and performance. However, Redman and Wilkinson (2006) observe that there is evidence of subjective manipulation of these 'fair' criteria. Performance, for example, relies on a good system of performance appraisal records; yet, as we discuss in more detail later, this is a contested area of resourcing practice.

Employee support may simply consist of allowing staff time off to look for work, which is a statutory requirement in the UK, and/ or help with job applications, personal finance planning, counselling and stress management. Outplacement services offer a more comprehensive way of providing employee support. But again, this is a costly operation and may not achieve the intended aims of 'considerate employer'. One experienced Outplacement Consultant reflects on the perverse reality he faces:

> when an organisation runs into difficulties and goes through the process of redundancy, the selection criteria for the initial phase often seems to seek out those with pre-existing and known performance issues, or those individuals who, for whatever reason, do not 'fit' with the current culture and ethos of the organisation. This is an activity that might be

termed 'deferred disciplinary action'. I believe that the management of the organisation often feels guilty about not dealing appropriately with these individuals earlier, and hence throws money at the problem by investing in outplacement services. If, however, the economic difficulties persist, then the second or subsequent phases of redundancy often hit those individuals that organisations actually would not want to lose. Unfortunately, at this time the economic difficulties are likely to be more significant, resulting in limited funds for employee support. It therefore seems to follow that outplacement support is often provided for those employees that the organisation probably wanted to part with anyway, and those arguably more valued employees at the later stages of the process are left unsupported.

Redundancies in construction

The construction management literature suggests that the sharp fluctuations in construction organisations' workload at times of downturn force larger scale redundancies in comparison to other industries, or that contractors retain inefficient surplus of staff in order to mitigate the difficulties in recruitment at upturn (Loosemore *et al.*, 2003). The figures for the current recession appear to support the latter (Office for National Statistics, 2010). Indeed, the construction industry seems to have done better than its major counterparts such as manufacturing or banking, finance and insurance (ibid.).

Together with the evidence about informal human resource planning (see discussion on managerial judgement above), this suggests that construction organisations have become well-adjusted to the uncertain and cyclical nature of the workloads and exhibit useful management techniques, such as flexible employment strategies, to deal with them. There may be few formal management systems in place (see for example Raidén *et al*, 2008, with regards to computerised human resource information systems); yet line managers' subjective decision-making practices seem valuable. This does not support the best practice hypothesis which calls for more formal (explicit) systems, nor does it guarantee consistency of practice or attention to the needs of the employees; but it highlights the embedded nature of empowerment throughout the industry. This aligns with ideology centred around high commitment model of HRM – an argument perhaps surprising in the context traditionally argued to focus on cost and operational issues.

Challenging best practice

As suggested at the start of the chapter, people resourcing activities are intended to ensure that an organisation has 'the right number of people with the right skills, behaviours and attitudes in the right place at the right time'. This phrase whose origins are somewhat obscure has been dismissed as being rather trite but repays analysis. It is implicit in this statement that what is 'right' for one organisation might not be 'right' for another; hence it supports the best fit rather than best practice approach to resourcing. However, the phrase also implies that there is a 'right' person, that there are 'right' skills, behaviours and attitudes and that there is a 'right' place and time. Moreover, it makes an assumption that these elements are indeed knowable and are interpreted and understood by all stakeholders in the same way. We now go onto discuss the factors which undermine assimilation and processing of information by managers to reach consistent decisions which will suggest that it is by no means certain this would be the case.

All people resourcing activities rely on people within an organisation (mainly managers) analysing information, drawing conclusions about what the information means and making decisions about what approach to take in relation to the issue. For each of the people resourcing activities to be carried out according to best practice, managers need information that is accurate and can make sense of, and their own decision making should be free of organisational politics and personal subjectivity. This is a fundamental flaw in notions of best practice.

The availability of accurate information will depend on the extent to which any organisation invests in the systems and procedures to collect relevant information, to ensure that it is up to date and is accessible to managers. An example of the type of information that would aid a manager to make decisions about promoting an employee or selecting him/her for redundancy would be records of performance appraisal grades and employee absence. However, in its latest absence management survey CIPD (2010c) note that 18% of organisations do not record absence data.

The availability of information aside; managers may not wish to make use of it, not know it is available or not be able to make sense of it and integrate it with other sources of information available to them on the same topic. Simon (1976) described this phenomenon as 'bounded rationality' and identified its impact on decision making. Applying this to resourcing activities, a useful example here is to consider the amount of information a manager will need to assimilate when assessing candidates at an assessment centre. Searle (2003) cites research which demonstrates that assessors frequently cannot process the behavioural information that they are observing rapidly enough and employ individual decision heuristics to arrive at a rating for each behavioural competency they are required to observe. In a typical assessment centre a manager–assessor might be required to observe two candidates simultaneously as they participate in a group discussion. In observing these candidates the manager–assessor may be required to assess what each candidate did and relate this to several behavioural competencies. Information overload in this case can often result in the assessment of a candidate's general performance and not rating specific behavioural competencies. Thus, if a manager's own preference is to 'get things done' they may rate a candidate who forces through decisions more highly than a candidate who adopts a more consultative approach, even if the competencies that they are rating are such behaviours as negotiating, influencing and teamworking. In effect the manager–assessor will see this as effective behaviour and give ratings to all the competencies under observation as if these were what were observed.

Finally there is the matter of whether managers can be relied upon to make decisions which are politically 'neutral' and free from subjectivity. Organisational power and politics will often influence managers to make decisions that are broadly in their own interests. When writing a job description or person specification they may inflate the academic requirements of the job in a bid to have more 'graduates' reporting to them or in human resource planning they may overestimate the numbers of employees required to increase the size of their departments and thereby control more organisational resources.

Subjective judgements arise from the perceptual biases which are hard-wired into every human being. Our perception of people is affected by our upbringing, education and experience and perceptual bias manifests itself through well-known mechanisms such as stereotyping, halo-horns, primacy-recency and attribution (see Martin, 2001). Resourcing activities are severely undermined by such perceptual biases: a manager may wrongly attribute to an employee blame for failing to meet a performance target; another manager might be biased in favour of candidates from the prestigious university he/she attended due to the halo bias. Person perception and perceptual bias are part of the responses we have retained from our evolutionary history and influence profoundly the entirety of resourcing activities.

We illuminate these issues via a Case Study and discussion as follows.

Case Study 9.1

Resourcing practice at GFT Construction Ltd

GFT Construction Ltd is a large UK-based contractor that operates in all main sectors of construction work: water engineering, rail and highways, commercial building, education and health, housing and industrial works. Their clients include both public and private developers. The company employs approximately 2,100 professional and managerial staff. Operative labour is mostly subcontracted. The organisation is well-known in the industry for its commitment to supporting the wellbeing and development of staff; some brand them a 'good employer.'

Their staffing effort is centred around team formation and deployment activities, as these are considered to be the most important of all the aspects considered under people resourcing:

'The real issue in construction is whether you can form good teams or not. This makes the difference between success and failure. Actually half of your long-term success is in the strength of your team.' (Senior operational manager)

Although 'getting it right' is considered extremely challenging due to the short-term time scales that apply to most construction projects, the organisation anchors all of the other people resourcing activities on this one central function. Finding suitable key people to head a project is the foremost concern in bringing together a project team. The selection criteria are based upon, in order of priority: employees availability, previous experience (ability), client preferences, individual's need for a particular job to gain experience or training, individual's personal aspirations (including their career management/ development needs), and the ability to devolve responsibilities (e.g. to develop and give experience to trainees on a project). This information is drawn together and discussed in divisional senior managers' weekly meetings. Senior managers' abilities are relied upon to fully understand the capabilities of their staff.

In addition to selecting the key personnel to head a project, ensuring a balance between the team members' strengths and weaknesses and their willingness to work together for a common aim is considered crucial:

'It is getting the balance right, the team members are not identical. This guy may be quite good at this, and the other may be weaker at that. It is like a piece of jigsaw; if it fits well into the situation with this guy then between them they can be quite strong.' (Operations director)

The team deployment activity is closely tied in with human resource planning, which is discussed in the same divisional senior managers' meetings. Indeed, team deployment is the key consideration in planning as project bidding dictates much of the decision-making. This is influenced by an organisation-wide strategic plan which is put forward by the board of directors with devolved targets on staff development and retention for each division to achieve. Divisional senior managers then reconcile the targets against their resourcing requirements with a view of ensuring that appropriately qualified and skilled staff are available and that there is a constant supply of new staff into their division. Not surprisingly therefore, recruitment and selection also connect closely with the team deployment activity.

The number and type of available vacancies is determined by workload (quantitatively) and nature of work. This means that even senior managers are initially brought in to run specific major projects, and then stay with the organisation after the completion of the particular project. Any shortfall in staffing is made up with temporary agency staff, and this method of staffing is used extensively throughout the organisation both short- and long-term.

Word-of-mouth recruitment and headhunting play a significant role in identifying new managers. The senior management team feel most comfortable using these techniques in ensuring that the new entrants have core interpersonal qualities such as a keenness to work as part of a team, assertiveness (but not aggression), the ability to fit in within the organisational culture and good communication skills. Technical competence, previous experience, personal skills and knowledge, and ambition are also seen as important characteristics of the managers that were likely to take the business forward. Behavioural interviewing is used to assess these criteria in selection interviews; divisional managing directors monitor the process for senior positions, but lower level vacancies are filled at the discretion of line managers at a project level. No formal training is provided for interviewers but the senior managers responsible for filling key positions appear confident that they know what is expected culturally. Culture is an important consideration as a strong sense of identity among the company staff was clearly identifiable and future recruits are sought to 'fit' the organisational culture.

An annual appraisal system provides an opportunity for discussing current job performance and personal development plans, and recording employees' aspirations and preferences. This includes employee self-assessment and an appraisal interview with their line manager. However, staff appraisal records are rarely considered in managers' decision-making process. This is mainly because only paper copies of the forms are kept. Also, sometimes appraisals are long overdue because of changing project commitments and an inability of managers to give meaningful feedback to personnel who they have worked with only short time.

The case study organisation appears to be using the best practice approach in so far as competencies have been identified and behavioural interviewing is used to assess these criteria in selection interviews. However, no formal training is provided for personnel involved in selection interviewing and word-of-mouth recruitment and selection is prevalent. Informality therefore clouds the effort. As acknowledged earlier in this chapter, the assessment and interpretation of behaviours is [still] mostly subjective. Furthermore, we found no evidence of job analysis in the recruitment process.

The many vacancies filled at the discretion of line managers and the search for organisational fit as a key principle in selection suggest that the 'social process' perspective to selection is more ubiquitous in this organisation than the systematic process related to best practice.

In terms of performance review, superficially the case organisation seems to support the best practice approach by offering a formal appraisal. Promisingly, both managers and employees have an input. However, because of the delays and since the appraisal records are rarely used in managerial decision-making the extent to which this process contributes to a sense of high commitment among staff is questionable. Also, while the organisation does collect end of project review data, this is not incorporated in the employees' appraisal process. Thus, an opportunity for collating feedback from multiple sources is lost.

It is evident from the case study that teamworking is central to its operations. All staffing activities (team deployment, human resource planning, and recruitment and selection) are focused on effective project team formation. The strength in this organisation's approach lies in their explicit consideration of the 'value added' elements of teamworking (such as team interaction) in managerial decision-making, which aligns closely with the high commitment approach. On the other hand, as noted in relation recruitment and selection above, the organisation is less organised in integrating systematic, organisation wide (possibly IT-enabled) technologies that enable objective decision-making.

The priorities of the case study organisation in selecting key personnel for project teams demonstrate the industry's tendency to focus on meeting immediate organisational/ project needs. They place employees' preferences and aspirations well down the priority list, although senior managers fully support learning and development and this is transparent in the extent to which development is considered in the latter three selection criteria. In practice, one manager reflects that:

> We probably do it from a gut feel really. I know the company, I have worked here for 27 years. You know people. There are newcomers, but I know the key players. It is all in here [pointing to his head]... (Raidén et al., 2009: 134).

This intuitive process opens up numerous possibilities for inconsistencies and subjective influence on the decisions.

As in many other construction companies, the case study organisation's internal labour market closely reflects Atkinson's (1984) flexible firm. At the centre is a core group of employees, the most valuable talent on permanent contracts of employment, that includes mostly managerial and professional employees. Alongside the core, the organisation 'employs' an extensive peripheral workforce (e.g. temporary agency staff and subcontracted operative labour force), that provide numerical flexibility. This allows effective management of change: staffing can be easily and quickly adjusted to match workload.

Another key strength evident in the case study above is their use of qualitative human resource planning in devolving strategic measures to manage staff development and retention at divisional level. Moreover, managerial judgement is exercised in 'getting it right,' which seems important. However, despite the strategic planning, at operational level the organisation tends to work on a rather short-term basis, taking the first of Taylor's (2010: 121–2) two suggested choices alluded to earlier: relying on managerial intuition, and reacting swiftly and decidedly as environmental change unfolds rather than developing contingency plans for a variety of possible circumstances.

Overall, informality, lack of training and integrated information, reliance on intuition and managerial judgement in planning are all factors which may lead to inconsistency of decision-making and allow subjective opinions to influence HR practice. One theme that comes across strongly in the case analysis is that the individual employees' aspirations are consistently overlooked as a consequence of lack of integrated information systems and the urgent nature of resourcing decision-making. While employees' aspirations are discussed at appraisal, they tend to get subordinated to the organisation's resourcing demands. Thus, while elements of best practice HRM were evident, the actual practice is not consistent with high commitment HRM (best practice) and is it is likely that this approach would undermine engagement initiatives.

Conclusion

In this chapter we have discussed HRM best practice and located people resourcing activities within its framework. Our discussions linked to the case study have challenged the extent to which organisations may be following the principles of best practice. The best practice perspective tends to promote a straightforward, 'one best way' of managerial decision-making that underplays the influence which organisational politics, errors in perception and the complex nature of relationships between individuals may have on resourcing activities. We suggest that people resourcing is a complex field of negotiation between informal practices that may be useful in engaging and empowering individuals, and formal mechanisms which attempt to minimise the limitations of human subjectivity and inability to process information in multiple locations simultaneously. It is our contention therefore that the best practice approach to resourcing offers only an illusion of equity, justice and consistency and that organisations must be constantly vigilant to ensure that these considerations are not neglected in the pursuit of organisational priorities. If employees suspect that in reality management practice is *ad hoc* and subjective they may become demoralised and organisations risk losing employee commitment and engagement.

References

Appelbaum, E., Bailey, T., Berg, P. and Kalleberg, A. (2000) *Manufacturing Competitive Advantage: The effects of high performance work systems*, New York: Cornell University Press.

Abdel-Wahab, M. S., Dainty, A. R. J., Ison, S. G., Bowen, P. and Hazlehurst, G. (2008) Trends of skills and productivity in the UK construction industry, *Engineering, Construction and Architectural Management*, Vol. 15, Issue 4, 372–82.

Atkinson, J. (1984) Manpower strategies for the flexible organisation, *Personnel Management*, August.

Baron, A., Clake, R., Turner, P. and Pass, S. (2010) *Workforce Planning Guide*, London: CIPD.

Biernat, M. and Fuegen, K. (2001) Shifting standards and the evaluation of competence: complexity in gender-based judgment and decision-making, *Journal of Social Issues*, Vol. 57, No. 4, 707–24.

Bloisi, W. (2007) *An Introduction to Human Resource Management*, London: Mcgraw-Hill.

Bradley, H. (2010) How to ... use assessment for redundancy, *People Management*, 14 October, 16–17.

Bramham, J. (1988) *Human Resource Planning*, Universities Press.

Chamorro-Premuzic, T. and Furnham, A. (2010) *The Psychology of Personnel Selection*, Cambridge: Cambridge University Press.

Chen, J., Yang, L., Chen, W. and Chang, C. (2008) Case-based allocation of onsite supervisory manpower for construction projects, *Construction Management and Economics*, 26: 803–12.

Cheng, M-I., Dainty, A. R. J. and Moore, D. R. (2005) What makes a good project manager? *Human Resource Management Journal*, Vol. 15, No. 1, 25–37.

CIPD (2003) *Competency Frameworks in UK Organisations*, London: Chartered Institute of Personnel and Development.

CIPD (2004) *People and Technology: is HR getting the best out of IT?* London: Chartered Institute of Personnel and Development.

CIPD (2005) *People Management and Technology: progress and potential*, London, Chartered Institute of Personnel and Development.

CIPD (2009) *Absence Management, Annual Survey Report*, London: CIPD.

CIPD (2010a) Performance Appraisal Factsheet, www.cipd.co.uk/subjects/perfmangmt/appfdbck/perfapp.htm?IsSrchRes=1 (accessed 24 August 2010).

CIPD (2010b) *Performance Management in Action: current trends and practice*, London: CIPD.

CIPD (2010c) *Absence Management, annual survey report*, London: CIPD.

Cook, M. (2004) *Personnel Selection: adding value through people* (4th edn), Chichester: Wiley.

Dainty, A. R. J. and Bagilhole, B. (2006) Women's and men's careers in the UK construction industry: a comparative analysis, Gale, A. W. and Davidson, M. J. (eds), *Managing Diversity and Equality in Construction*, Abingdon: Taylor & Francis, 98–112.

Dainty, A. R. J., Cheng, M-I. and Moore, D. (2005) A comparison of the behavioral competencies of client-focused and production-focused project managers in the construction sector, *Project Management Journal*, Vol. 36, No. 1, 39–48.

Donkin, R. (2010a) *The Future of Work*, New York: Palgrave Macmillan.

Donkin, R. (2010b) The world we live in, *People Management*, 11 February, 22–24.

Druker, J. and White, G. (1996) *Managing People in Construction*, London: IPD

Gallup (2010) Employee engagement, a leading indicator of financial performance, www.gallup.com/consulting/52/employee-engagement.aspx (accessed 30 October 2010).

Huselid, M. A. (1995) The impact of human resource management practices on turnover, productivity, and corporate financial performance, *Academy of Management Journal*, Vol. 38, No. 3, 635–72.

Iles, P. (1999) *Managing Staff Selection and Assessment*, Buckingham: Open University Press.

Kahn, W. (1990) Psychological conditions of personal engagement and disengagement at work, *Academy of Management Journal*, Vol. 33, No 4, 692–724.

Koskenvesa, A., Koskela, L., Tolonen, T. and Sahlstedt, S. (2010) Waste and labour productivity in production planning, *18th Annual Conference of the International Group for Lean Construction*, Haifa, Israel, 14–16 July, 477–86.

Latham, G., Sulsky, L. M. And Macdonald, H. (2007) Performance Management, Boxall, P., Purcell, J. and Wright, P. (eds), *Oxford Handbook of Human Resource Management*, Oxford: Oxford University Press.

Lawler, E. E. and Mohrman, S. A. (2003) HR as a strategic partner: what does it take to make it happen? *Human Resource Planning*, Vol. 26, No. 3, 15–29.

Legge, K. (1995) *Human Resource Management: Rhetoric and realities,* London: MacMillan.

Lockyer, C. and Scholarios, D. (2007) The 'rain dance' of selection in construction: rationality as ritual and the logic of informality, *Personnel Review*, Vol. 36 Issue 4, 528–48.

Loosemore, M., Dainty, A. R. J. and Lingard, H. (2003) *Human Resource Management in Construction Projects*, London: Spon Press.

Longenecker, C. (1997) Why managerial performance appraisals are ineffective: causes and lessons, *Career Development International*, Vol. 2, No. 5.

Lowry, D. (2002) Performance Management, In Leopold, J. (ed), *Human Resources in Organisations*, Harlow: Financial Times Prentice Hall, 129–65.

Newell, S. (2006) Selection and assessment, Redman, T. and Wilkinson, A. (eds), *Contemporary Human Resource Management; Text and cases* (2nd edn), Harlow: Financial Times Prentice Hall, 65–98.

Newell, S. and Rice, C. (1999) Assessment, selection and evaluation: problems and pitfalls, Leopold, J., Harris, L. and Watson, T. (eds), *Strategic Human Resourcing, Principles, practices and perspectives,* London: Financial Times 129–65.

Ng, S. T. Skitmore, R. M. and Sharma, T. (2001) Towards a human resource information system for Australian construction companies, *Engineering, Construction and Architectural Management*, Vol. 8, No 4, 238–49.

Marchington, M. and Wilkinson, A. (2008) *Human Resource Management at Work* (4th edn), London: CIPD.

Martin, J. (2001) *Organizational Behaviour* (2nd edn), London, Thomson.

Martin, G., Reddington, M. and Kneafsey, M. B. (2009) *Web 2.0 and Human Resource Management: 'groundswell' or hype?* London: CIPD.

Martin, M., Whiting, F. and Jackson, T. (2010) *Human Resource Practice* (5th edn), London: CIPD.

McGregor, D. (1957) An uneasy look at performance appraisal, *Harvard Business Review*, Vol. 35, No 3, 89–94.

McLeod, D. and Clarke, N. (2009) *Engaging for Success: enhancing performance through employee engagement*, www.bis.gov.uk/policies/employment-matters/strategies/employee-engagement (accessed 1 September 2010).

Office for National Statistics (2010) *Economic and Labour Market Review May 2010*: www.statistics.gov.uk/elmr/05_10/2.asp (accessed 17 May 2010).

Pfeffer, J. (1998) *The Human Equation: Building Profits by Putting People First*, Boston: Harvard Business School Press.

Phillips, T. (2005) *After 7/7, Sleepwalking to Segregation*, www.humanities.manchester.ac.uk/socialchange/research/social-change/summer-workshops/documents/sleepwalking.pdf (accessed 20 May 2011).

Pilbeam, S. and Corbridge, M. (2010) People Resourcing and Talent Planning: HRM in Practice (4th edn), Harlow: Financial Times Press.

Pollert, A. (1991) *Farewell to Flexibility?* Oxford: Blackwell.

Purcell, J. (1999) Best practice and best fit: chimera or cul-de-sac? *Human Resource Management Journal*, Vol. 9, Issue 3, 26–41.

Purcell, J., Kinnie, K., Hutchinson, S., Rayton, B. and Swart, J. (2003) *Understanding the People and Performance Link; Unlocking the Black Box*, London: CIPD.

Raidén, A. B., Dainty, A. R. J. and Neale, R. H. (2001) Human resource information systems in construction: are their capabilities fully exploited? *17th Annual ARCOM Conference*, University of Salford, Salford, UK, 5–7 September.

Raidén, A. B., Dainty, A. R. J. and Neale, R. H. (2006) Balancing employee needs, project requirements and organisational priorities in team deployment, *Construction Management and Economics*, Vol. 24, No. 8, 883–95.

Raidén, A. B., Williams, H. and Dainty, A. R. J. (2008) Human resource information systems in construction – a review seven years on, *24th Annual ARCOM Conference*, University of Glamorgan, Cardiff, UK, 1–3 September 2008.

Raidén, A., Dainty, A. and Neale, R. (2009) *Employee Resourcing in Construction*, Abingdon: Taylor & Francis.

Rawls, J.B. (1999) *A Theory of Justice*, Oxford: Oxford University Press.

Redman, T. (2006) Performance appraisal, Redman, T. and Wilkinson, A. (eds), *Contemporary Human Resource Management; Text and cases* (2nd edn), Harlow: Financial Times Prentice Hall, 153–87.

Redman, T. and Wilkinson, A. (2006) *Contemporary Human Resource Management; Text and cases* (2nd edn), Harlow: Financial Times Prentice Hall.

Robinson, D., Perryman, S. and Hayday, S. (2004) *The Drivers of Employee Engagement*, Brighton: Institute for Employment Studies.

Saks, A. M. (2006) Antecedents and consequences of employee engagement, *Journal of Managerial Psychology*, Vol. 21, No 7, 600–19.

Scarbrough, H. (1999) The Management of Knowledge Workers, Currie, W. L. and Galliers, B. (eds), *Rethinking Management Information Systems*, Oxford, Oxford University Press.

Scarborough, H. and Carter, C. (2000) *Investigating Knowledge Management*, London: CIPD.

Searle, R., (2003) *Selection and Recruitment, a Critical Text,* Milton Keynes: Open University.

Simon, H. (1976) *Administrative Behavior* (3rd edn), New York: The Free Press.

Sparrow, P. R. and Bognanno, M. (1993) Competency requirement forecasting: issues for international selection and assessment, *International Journal of Selection and Assessment*, Vol. 1, No. 1, 50–58.

Syedain, H. (2010) Workforce Planning: A force for good, *People Management*, 3 June, 24.

Tansley, C., Newell, S. and Williams, H. (2001) Effecting HRM-style practices through an integrated human resource information system: an e-greenfield site? *Personnel Review*, Vol. 30, Issue 3, 351–71.

Taylor, S. (2010) *Resourcing and Talent Management* (5th edn), London: CIPD.

Torrington, D., Hall, L. And Taylor, S. (2002) *Human Resource Management*, 5th edn, Harlow: Financial Times Prentice Hall.

Walsh, D. and Lugsden, E. (2009) Parting company: the strategic responsibility of exit management, Leopold, J. and Harris. L. (eds), *The Strategic Managing of Human Resources* (2nd edn), Harlow: Financial Times Prentice Hall, 250–88.

Watson, T. (2002) *Organising and Managing Work; organisational, managerial and strategic behaviour in theory and practice*, Harlow: Financial Times Prentice Hall.

Williams, H., Tansley, C. and Foster, C. (2009) HRIS Project Teams Skills and Knowledge: A human capital analysis, Bondarouk, T., Ruël, H., Guiderdoni-Jourdain, K. and Oiry, E. (eds), *Handbook of Research on E-Transformation and Human Resources Management Technologies: Organizational outcomes and challenges*, London: Information Science Reference, 135–52.

Wong, J. M. W., Chan, A. P. C. and Chiang, Y. H. (2011) Construction Manpower Demand Forecasting: A comparative study of univariate time series, multiple regression and econometric modelling techniques, *Engineering, Construction and Architectural Management*, Vol. 18, No. 1, 7–29.

Wyatt, W. (2008a) Companies set up workforce planning activity in response to economic slowdown, 4 December, www.hr.com (accessed 26 January 2009).

Wyatt, W. (2008b) One in five firms unprepared for the economic downturn, *Personnel Today*, 28 October, www.personneltoday.com/articles/2008/10/29/48161/one-in-five-firms-unprepared-for-the-economic-downturn.html (accessed 26 January 2009).

10 Reward management in construction
Janet Druker

Introduction

This chapter considers the management of reward – including the management of pay and benefits at work – as one of the key levers in the delivery of strategic human resource management (Storey, 1993). The objective of the chapter is to open up further discussion of reward management within the construction industry. It is a topic that has been relatively neglected within the construction literature yet it is a vital ingredient in project delivery, in attracting and retaining high calibre employees and in ensuring that a committed and skilled workforce remains in place for today's project and is available as a resource for the future.

First, we review the reward literature and 'new pay' theories. We also consider the externalisation of labour engagement – a point that is especially relevant to a discussion of pay and benefits because of the ways in which the construction industry draws on the informal economy – with the prevalence of self-employment in the UK construction sector providing a vivid illustration. Second, we consider the factors influencing pay determination, including legislation and collective bargaining and referring to debates concerning national cultural factors. Third, we turn to a discussion of the construction industry in the UK and after a brief comment on employment structures and organisational change, there follows a discussion of pay and benefits as evidenced from published sources, comparing the changes affecting different occupational groups. A case study follows. The chapter concludes with a discussion of the relevance of HR and organisational theories to the diversity of reward practices within the industry.

Theory and practice

The rewards of daily working life are often taken to be concerned with pay, since it is pay that provides the return for work that has been undertaken and it is inherently important in the exchange for goods and services. In addition to its exchange value, pay may also, through grading structures and internal relativities, suggest social positioning, both in the workplace and more widely. Pay determination and pay structures vary significantly across industries and national boundaries but total earnings are likely to encompass a basic wage, to which may be added variable pay including bonus or incentive schemes. Overtime pay may be awarded for extra hours worked over and above the stipulated norm. In some cases the entirety of the earnings package may be dependent upon performance, or payment will be made on the basis of completion of a particular task. Additional benefits play a part in reward too and these could include provision for holidays, health and life insurance, sickness, parental leave and pension arrangements. Such benefits may be sizeable constituents in the overall reward package. These factors, together or separately, may influence motivation, affecting employee engagement and performance and, where there is choice, influencing the likelihood that the individual will remain with the organisation.

Job satisfaction may be a component within the experience of work and the notion of 'reward' is bound up with complex psychological and social factors as well as with economic issues. There is a significant literature – particularly American literature – concerned with the question of what motivates people at work. This may both reflect but also reinforce employer expectations and decisions. The 'scientific management' theories of Fredrick Winslow Taylor (1913) emphasised the importance of 'economic man', motivated by financial reward as a return for the effort involved at work. With effective work systems and appropriately geared incentive schemes, workers could, in Taylor's view, achieve a 'win-win' situation by maximising their income as they achieved the levels of productivity anticipated by management. This approach has remained an influential strand in management thinking, notwithstanding the changes of the last century. It tends to be associated both with payment by results schemes and with individual performance-related pay.

The human relations school, with its origins in research conducted by Mayo (1933 cited in Steers, *et al.*, 1996: 28) and by Roethlisberger and Dickson (1939, cited in Steers, *et al.*, 1996: 28) leant more weight to the social and group dynamics involved at the workplace, with group norms and personal recognition according a feeling of belonging, which in turn contributed to enhancing performance. The motivational philosophy here was concerned with ensuring that workers felt that their role was relevant and important. Communication was seen as having a particularly significant part to play and, where financial rewards were considered, there was greater emphasis on group rewards. From the very different perspectives offered by Taylor and by Elton Mayo we can see something of the variation in views about what is valued in and through the experience of work. In reviewing the literature and in seeking to understand the different orthodoxies espoused by managers, it is important to appreciate, as Wilfred Brown made clear some 50 years ago, that there is a complex interplay of factors involved in establishing the most effective reward management system and that there is no single, easy solution to the challenges involved in doing so (Brown, 1962).

From the employee's perspective, perceptions of the opportunities and benefits of different occupations shape individual decisions in respect of training and career choice. However following labour market entry it is the lived experience that governs job satisfaction and sets up a 'psychological contract' between employer and worker (Guest and Conway, 2001). Rewards are often said to be of two different types. First, there are intrinsic rewards, those derived from the nature of the work itself and the satisfaction it may bring and from the work environment. Hence it can be argued that reward is not simply about pay and benefits. Second, there are extrinsic rewards – elements that are concerned with the tangible or transactional return – with salary, incentive pay and with benefits (Perkins and White, 2008: 8–9). Within these considerations the management of pay and benefits (sometimes termed 'compensation' in the American literature) looms large.

The relationship between the employing organisation and the individual to be employed is founded on a promise – the promise of wages in return for work. The nature of the exchange requires definition with regard to the tasks that are to be undertaken, the results that are to be achieved and the price that is to be paid. Three points of reference have been identified here 1) performance or output; 2) the job itself and the time that is worked; and 3) the personal qualities of individual employees and the particular skills that they may have (White, 2009: 25). In addressing these, pay determination requires attention to what is felt to be fair – both by reference to internal comparisons and also by reference to the external labour market. This is a matter of increasing public concern as publicity concerning bonus payments made to senior executives in the finance sector has highlighted massive disparities. The challenge with respect to internal equity tends to be greater in larger organisations where the creation of a grading structure may set out the hierarchy of jobs against which pay can be set. This is not in itself the basis for pricing jobs – which will be measured against others in the external labour market (White, 2009: 25). Nonetheless a 'Fair Pay code' has been advocated for the public sector in the UK and a recent report has pointed to the need for principles of greater transparency in senior pay and pay multiples (Hutton, 2011).

The 'new pay' literature about reward management, reflecting on the internal labour market, points to a logic which connects business strategy to management practice. Edward E. Lawler the Third, one of the notable writers in this field, commented (2000: 249) that 'Many organisations have good strategies but only a few implement them effectively' and that 'The reward system is an important determinant of how effectively a business strategy is implemented.' Premising his views on the assertion that people are central to organisational success, he pointed to two key components that should be addressed. First, there is the development of organisational competence; and second, the creation of a motivational strategy through which results can be delivered. Other new pay writers, following this lead, argue that convincing employees to deliver against organisational objectives means ensuring that people understand what constitutes effective performance and that individuals and teams are encouraged to perform well (Schuster and Zingheim, 1992). Policies on pay and benefits need to be fully and effectively integrated with other aspects of organisational design and practice, they say. Lawler refers to the 'reward system architecture' which must take account of payment systems, bonuses, team or group working as well as of issues such as market position, equity and hierarchy. On the one hand it is essential to manage costs, including financial and risk controls whilst on the other, decisions must also take account of the perceived equity of financial returns and the practical benefits of work. Without an effective alignment between these issues, organisations may be at risk. Moreover, internal reward management practices should be effectively communicated so that individuals both understand what is required of them and feel valued and motivated. So it is in part a question of what is done and in part a consequence of how it is done – how decisions are taken and communicated.

Organisational boundaries

The new pay literature is essentially normative in approach – presenting the desirability of a holistic and strategic approach to the management of reward. It is based on the perspective of a single organisation and does not seek to question the notion of what constitutes the 'organisation'. In reviewing reward practices, account must be taken of alternative approaches to the management of pay and benefits, concerned essentially with use of the external as opposed to the internal labour market. Rather than viewing employer decision-taking through the lens of a single organisation wherein employers seek to retain key skills and to reinvest in the workforce, employer decision-taking may be understood within the more complex framework of inter-organisational relations. Drawing on segmentation theory, it is argued that many organisations in the construction industry will seek to 'downsize and redistribute' activities so that organisational boundaries are blurred (Grimshaw *et al.*, 2005a). In this way the formalised structures of the larger organisation and the scope for comparison and reference to internal relativities can be avoided. Studies of HR issues are often undertaken from the perspective of a single organisation but Rubery and Earnshaw (2005) suggest that employers may undertake a process of externalisation so that employment structures are fragmented, making trade union organisation less common. Pay levels and relativities are less open to challenge within the complexities of inter-organisational interaction. This chapter argues that this theoretical perspective is relevant to an understanding of reward management within the construction industry in the UK, where self-employment is commonplace.

In considering the 'architecture' of reward management, it is important to note that employers have significant choices in framing the approach that they take. It is true that they are subject to legal requirements in terms of the management of the employment relationship. It is true too that these statutory obligations extend to some aspects of the management of pay and benefits and in the UK context legislation on these issues has become more extensive in recent years, because of both national legislation and European law. Nonetheless employers in the UK (as in many other countries outside of Europe) retain significant freedom in determining how they will manage within their own organisations and it is to this issue that we now turn.

Pay determination

Employers determine the location and the nature of their business and do so with a view to optimising opportunities and removing constraints. In itself this variation may shape employer behaviour and decisions of multi-national enterprises for example through a choice to invest in a low wage economy or one which is less regulated, as a means to retaining control of costs and of work organisation.

Pay and associated benefits are internationally significant as motivating factors in relation to work although their precise meaning will differ according to national and cultural context (Sparrow, 2009). The national context is important because of the variation in the legislative, collective and social security arrangements across national boundaries. Legislative provision is important in setting the framework within which employer decision-taking is framed. So for example, the law may set standards on minimum pay levels – as it does in the US, in France and in the UK. There may be legislation governing working hours (e.g. the 1993 European Working Time Directive). The principle of equal pay is to be found in regulation in many developed countries and in the European Union it was an integral feature of the founding Treaty of Rome, revised and amended subsequently in the light of legal judgment and experience (Leat, 2003: 234-5). State benefits and employer or sectoral provision may be interconnected, since where the state fails to provide, employers may be expected to do so. For example, in the US, where there has been no comprehensive health care provision (other than Medicare), employer-supported health insurance is a valuable work-based benefit and those without work may be deprived of health cover (Mishel *et al.*, 2009: 353). Where benefits are provided by the state – as retirement pensions are in many OECD countries for example – there may be less pressure on employers to provide through occupational pension funds (Taylor, 2009: 192). Where governments seek to minimise the cost of state benefits they may require employers to make a greater contribution. In the UK, where successive governments have been seeking to contain public expenditure, there is an attempt to ensure greater employer contribution from employers to pensions. The 2008 Pensions Act requires that from 2012 employees be enrolled into a pension scheme to which the employer must make a minimum contribution (Pensions Act 2008).

In addition to state provision, there may be employer, firm-level or sectoral arrangements deriving from institutional interest representation and collective bargaining between trade unions and employers' representatives. Stronger unions, higher minimum wages and more generous social benefits are more commonly found within Europe where sectoral social partners may provide for sickness and welfare benefits (Mishel *et al.*, 2009), even though European countries differ very significantly in the form that such arrangements may take. There is some co-ordination and mutual support through social partnerships at European level but as yet no provision for European-wide collective bargaining. In some cases legal provision may underpin and support or extend decisions that are reached through collective bargaining – for example in Germany. Trade union membership and support for employers' associations has been diminishing in many developed countries (e.g. US, Australia and Western Europe) over the last quarter of a century and the influence and reach of collective bargaining is less extensive where this is the case (see further Chapter 8).

So where collective bargaining does not have legal underpinning or universal application and where there is no formal recognition by the employer of trade union representation and rights to bargain collectively, the employer has wide discretion in terms of the process of pay determination and in terms of pay levels and adjustments that may result. This is the case in most private sector enterprise in the US, in the UK and in many African and Asian countries, subject only to the implications of legislation.

Standing (1997) suggested that attention should be given to the 'social wage' – the total package from all sources – comprising the money wage and other benefits paid by the employer or through occupational provision, the benefits received from the state and support that may normally be anticipated through kinship or community support (which might be higher through family support networks in south-east Asia but lower in Europe). This is a helpful reference point, recognising the need to understand reward practices within a particular socio-political context.

Cultural factors may influence employer decisions and also employee preferences, although care is needed in the meaning that is attributed to notions of 'national culture' and the implications that this may have for reward practice (Sparrow, 2009). Labour migration across national boundaries as well as the multicultural nature of workplaces both in developed and developing economies combine to make this a difficult issue to unravel. Sparrow, (2009: 233–57), has reviewed recent research in this field. Whilst the literature (especially where it draws on Hofstede's cultural categories) suggests long-standing and stable cultural influences on reward decisions, Sparrow argues that employee behaviours cannot readily be predicted in this way and that employee preferences may change over time. Whilst recognising the breadth of research over the last decade he suggests a gap in understanding particularly with respect to the impact of culture on different types of reward, arguing that ultimately the evidence suggests that institutional factors weigh more heavily than national culture.

Context for reward in the UK construction industry

An understanding of reward management has to be grounded in the context of developments in the industry overall. Over the last 25 years, the largest construction organisations in the UK have reduced direct employment in favour of networks of activity through which construction processes and projects are managed (Dainty *et al.*, 2004). This type of reorganisation has taken place in many developed countries (McGrath-Champ and Rosewarne, 2009) and so it is relevant to consider the UK as a case study where there is a long contractual chain. In the UK context structures have been extensively 'de-layered', leading to substantial increase in self-employment. The industry was buoyant between 1997 and 2007 with a rise both in employment (from 1,107,000 employees in 1997 to 1,292,000 in 2007) and in self-employment (from 664,000 in 1997 to 953,000 by 2007; ONS, 2009a). The competitive challenge has been heightened since 2007 by post-recessionary risk, by cutbacks in capital expenditure within the public sector and by pressures on borrowing. This uncertainty encourages employers to think in terms of risk management and cost control.

The cost of labour is a key component within the contract price and the subject of careful calculation in tendering. Whether employed or self-employed, labour cost is a factor to be managed yet the commercial contract is unlikely to set specific requirements with reference to the form of labour engagement, the price of labour or the conditions attached to the employment relationship within the main contractor organisation or within the associated sub-contractual relations.

As noted above, the psychological contract may hinge upon intrinsic rewards as well as on pay and benefits. In interviews with construction managers conducted to support the Case Study in this chapter, individuals suggested that, at its best it may offer scope for creativity, innovation and a feeling of long-term impact on the environment 'the desire to make something that will last' (Project Director). The greater autonomy and variety of work and work situations over time may also be valued. As one project manager noted, 'You're not stuck in the same train, going to the same office with the same people'. The construction process offers managers the prospects of reasonable financial returns for the hardworking and innovative individual and continues to attract young men entering the labour market into a varied range of careers.

These intrinsic rewards do not have a uniform appeal and pay can prove to be a strong cause for satisfaction/ dissatisfaction – with the employer, with the industry or both. The importance of pay and pay relativities has been illustrated in the UK construction sector by a study of the psychological expectations of construction project managers in the UK. It pointed to the significance in terms of perception, of the pay differential between craft workers and first line supervisors and their line managers – and the damage caused by its erosion (Dainty *et al.*, 2004: 36). Another study – a survey of architects – revealed that the rate of pay was the factor most commonly leading to dissatisfaction with 41 per cent of respondents expressing dissatisfaction with the rate of pay, as compared with 35 per cent for the way the practice is managed and 34 per cent with opportunities for promotion (Sang *et al.*, 2009). The consequences in such cases may be dysfunctional for the firm since employees may leave the organisation for work elsewhere – or in remaining they may deliver a lesser performance than they have the capacity to deliver and one that is sub-optimal from the perspective of the employing organisation.

The question of equal pay for women is a difficult one in an industry that remains male-dominated because women's employment remains so low that reliable statistical comparison is impracticable. The composition of the labour force continues to be highly gendered (Dainty *et al.*, 2001). The construction industry provided work for 5.5 per cent of the employed population in the second quarter of 2010, but this was made up of 9.1 per cent of men (down from 10.4% a year earlier) but only 1.7 per cent of women (ONS, 2010). These industry-based figures probably understate the real gender divide since women are often found in traditional female roles in administration and are scarcely to be found in craft occupations. This relative absence of women means that there has been little pressure for equal treatment in terms of pay and benefits but it also means that the presence of women in the workforce is still too low for the statistics to yield any insights in terms of gender pay comparisons.

From 1919 until 2009 the pay of building and civil engineering operatives was determined (in theory at least) through national collective bargaining - a voluntary collective process in the UK through which pay and conditions were agreed on a multi-employer basis. Other trades had separate collective agreements set up during this period – for example in engineering construction and in electrical contracting (Korczynski, M., 1997; Gospel and Druker, 1998). The purpose was essentially to take labour out of competition within the competitive contracting process. Employers or their representatives (typically employers' associations) and employees and their representatives (trade unions) played a part in the national determination of pay and conditions agreeing hourly pay rates and allowances, working hours, holiday arrangements and associated conditions that, for many years, were applied to operatives (e.g. Construction Industry Joint Council, 2006). These arrangements in construction have, at least on paper, outlasted comparable agreements in other sectors which, typically, were dismantled in the last decades of the twentieth century. In construction, their importance has diminished although it has not entirely disappeared. Whilst the formal structures of national collective bargaining in construction were retained until 2009, collective agreements were often ignored and employers reserved the right to make unilateral pay arrangements at site level. In this way the appearance of national bargaining was accompanied by unilateral employer decision-taking, with self-employed labour widely used and normally unchallenged.

Collective bargaining coverage in the construction industry was reported by the authoritative (and most recent) Workplace Employment Relations Survey (WERS) as standing at 24 per cent in 2004 by workplace, down from 32 per cent in 1998, revealing an annual reduction of 4.2 per cent (Kersley *et al.*, 2006: 187). This compared with figures for all workplaces of 38 per cent in 1998, declining to 35 per cent by 2004 – a decline of 1.1 per cent per year. This trend has undoubtedly continued since the last WERS survey. Employers may continue to refer to collective agreements as a benchmark for labour costs and as a means of mitigating risks of labour organisation and potential disruption (Druker, 2007). However the degree of attention given to the terms of collective agreements in practice is not high. It will vary from one trade to another, and one agreement to another, being more widespread within electrical contracting (through the Electrical Contracting Joint Industry Board) and less commonplace in construction (explaining what may seem to be a high figure of 24% for the industry as a whole in 2004). The same WERS survey indicated that only 5% of employees in construction workplaces with fewer than 100 employees (that is the majority of workplaces) were 'covered' by collective bargaining. In practice being 'covered' might mean simply that the level of settlement each year is applied to pay rates (even where those rates are not the same as those prescribed in the national collective agreement) so that higher level settlements (e.g. in 2004–05) have tended to have some impact on actual earnings, even where employers were not committed to implementation of other aspects of the agreement. There has been no agreed pay settlement in construction since 2008 and the employers' association, the Construction Confederation, was dismantled in 2009. This raises further questions about the influence of collective bargaining on reward practice – a topic to which we will return when we consider the pay and conditions of hourly paid workers.

Whilst legal standards and collective bargaining agreements set some benchmarks for employers in the construction sector, they impose few real constraints and employers are largely free to determine the values and the practices which will inform their own organisational approach to the management of reward. It is to this issue that we now turn.

Reward within and across organisations

The first impression is of the sheer variety of reward practices that characterise the industry. Senior and strategic managers, project and contracts managers, technical and specialist advisors and support staff are employed on a salaried basis – paid an annual salary, usually paid monthly by credit transfer. Notice periods are likely to be at least one month. Salaries may be augmented by an additional performance-related bonus payment. Operatives, both skilled and unskilled, who carry through the main construction activities are paid 'wages' calculated on an hourly or daily basis. Typically they are lower paid and lower paid workers are more likely to receive their pay weekly or fortnightly and to have briefer notice periods (Druker, 2009: 104). They may be eligible for some form of performance bonus too. Other trades specialists and operatives engaged through contractors and sub-contractors, deployed to deliver the project and to ensure that specialist needs are met are often engaged on a 'self-employed' status and paid a day-rate, paid for the task or a combination involving payment for the time and the task. Their engagement may be less formal and their wage levels and conditions of work will not be included in the official statistics about 'employees'.

The contrasts in reward practice are embedded within the framework of labour engagement so it is important to consider the differences between these arrangements. Employment status is important in defining the form of payment and associated expectations, the level of payment and the nature of associated work benefits that may be received by individuals within each of the different groups.

It might be argued that, if 'new pay' theories outlined above have any meaning, those workers whose contribution to the delivery of business strategy is most significant should benefit from the highest rewards and the greatest security. This in turn raises the question of who is perceived by decision-takers to make the greatest or most significant contribution. Put another way, who is it who is indispensible to organisational success? Do project managers, for example, receive payment in line with the key position that they are said to hold? And if they do, what does this mean for other employees – and for the self-employed – within the industry?

Reward strategies, levels and composition of pay within the sector

Salaried staff

There are significant contrasts between the employment contract that might apply to professionals and managerial staff, whose pay is expressed in terms of an annual salary, and the position of operatives within the industry, whose pay (when employed directly) is often calculated according to an hourly or weekly norm. The principles governing pay for construction managers, project managers, site managers, section managers, engineering construction manager(s) and other professional, specialist or support staff are rooted in annualised systems. Payment for the individual is expressed in terms of an annual salary with an expectation that the individual will work the hours required to get the job done. Whereas for waged workers, increased working hours may be compensated by increased hourly payments, the typical salaried arrangement makes no provision for overtime and so under contract pressures, the more extensive hours of work are completed as part of an internalised discipline, through commitment to the organisation – and perhaps in the expectation of an annual bonus – rather than because of financial remuneration reflecting additional hours worked. This means that in the busiest periods – when directly-employed waged workers might be receiving additional, sometimes premium pay for overtime, those who are salaried are working longer hours for no additional recompense. The maximisation of pay in the short term may be less important in terms of overall satisfaction than some other facets of the employment contract, notably the prospect of continuity in earnings and the possibility of progression within the organisation. But long working hours take their toll on those in management grades and in reviewing the position of project managers, Dainty *et al.* (2004: 37) reported a 'trade off' whereby they might be prepared to accept lower value packages in exchange for the advantage of being able to work close to home (in effect shortening the hours committed to work).

It is interesting and revealing to ask how a salary is determined, a question that was posed in the course of research for the case study below and deserves further and more extensive research. In larger organisations there may be some kind of internalised grading structure which positions individuals within the hierarchy of the firm, with pay enhancement depending on professional standing, experience, performance and perceptions of overall ability. Advice may be sought from professional pay surveys which benchmark and compare pay levels for different occupations (see e.g. Hays, 2010). Of course it is the larger firms that are most likely to have human resource managers who can advise and support innovation but even in smaller organisations, there may be a structured approach to internal relativities (albeit less formally articulated). In setting pay levels, decision-takers will have an eye to the ability of the firm to pay. They will also be aware of competitor practices and of the likelihood of their more valuable staff being 'poached' by another organisation.

Good practice from other sectors – evident in some of the larger firms – points to broad-banded pay and transparency in reward systems. Each job profile is assigned to a job 'family' which in turn is positioned within a broad band. Each broad band has agreed levels of pay, benefits and reward. The job profile outlines the main responsibilities of the role. Job profiles are specified in this way for every managerial and professional role – from trainee through to the most senior positions. There is the possibility of progression within the band – or in the event of more significant progression movement up into the next band.

Broad banding is a more difficult system to control than a traditional multi-graded scheme and requires training for line managers and effective communication with employees if it is to be effective (Armstrong and Murlis, 2007: 619). Unilateral pay determination may well be more commonplace with salary increases awarded according to the perception and preference of senior managers within the firm on the basis that they understand the industry – or at least their own part of it – and the role of the individual managers and professionals who work for them. Dainty *et al.* argue that project managers have become the key human resource for modern contracting (2004: 34) with these changes accompanied by reports of increased salary levels and incentive payments. This is an interesting point of reference as we consider the experience of different occupational groups within the discussion that follows.

Information about the level and composition of pay has been collected and published by the Office for National Statistics in the Annual Survey of Hours and Earnings (ASHE) since 2004. In order to understand the relative position of different groups we consider below the position of construction employees relative to other comparable groups – comparing construction managers with other managers, construction professionals with other professionals and construction trades with other skilled trades.

Table 10.1, extracted from the ASHE 2004 and 2009, compares the pay of managers and professional occupations within the construction industry with managers and senior officials in all sectors. It also looks at the changing picture over the five-year period from 2004 to 2009 and presents a particularly interesting picture in terms of improvements to the pay of construction managers.

Table 10.1 Annual Survey of Hours and Earnings (ASHE) 2004–09

	2004		2009		% Change 04–09	
Description	*GWP* Median*	*GWP Mean*	*GWP Median*	*GWP Mean*	*GWP Median*	*GWP Mean*
	£	£	£	£	£	£
Managers and senior officials	585.4	706.8	681.0	820.0	16.3	16.0
Managers in construction	604.1	648.6	765.2	865.9	26.7	33.5
Professional occupations	575.8	616	652.0	710.4	13.2	15.3
Civil engineers	586.4	598.5	667.9	729.7	13.9	21.9
Architects	599.1	650.6	689.9	761.7	15.2	17.0
Town planners	573.3	577.9	655.6	697.9	14.4	15.6
Quantity surveyors	576.6	583.8	707.7	716.9	22.8	22.8
Chartered surveyors (not QSs)	575.8	648.4	669.5	707.5	16.3	9.1

Data based on Office for National Statistics Annual Survey of Hours and Earnings (ASHE), 2004 & 2009b. *GWP = Gross Weekly Pay.

Table 10.1 shows that in 2004, the median weekly pay of managers and professionals in the construction sector was above managers and senior officials more generally whilst the mean for construction in 2004 tended to lag behind the mean for all managers. This reflects the fact that the highest level of pay for managers and senior officials in all sectors was significantly ahead of those for construction, with upper decile earnings standing at £1,231.4 weekly (all managers) and £926.1 (construction) respectively.

Between 2004 and 2009, construction managers saw a significant improvement to their position, with a 26.7 per cent increase in median weekly pay and a 33 per cent increase in the mean. This contrasted with a 16 per cent increase on both counts for all managers and senior officials and took both the median and the mean for construction above the general level. Whilst upper decile gross weekly pay for all managers stood at £1444.60, continuing to remain ahead of construction at £139.50, the gap was much narrower than five years previously, both in absolute and relative terms. A similar picture is shown for construction professionals too, although the gains over the period are less marked than for construction managers.

The changes in weekly pay for these groups reflect three things. First, in a volatile industry, earnings in periods of prosperity may rise to ensure retention and to compensate for an insecure work environment and the risk of lay-off in a period of downturn. Second, the complexity of externalised labour and a long contractual chain carries a business cost in terms of higher pay for managers. Extensive sub-contracting and the transactional challenges that are posed for project delivery require high-level management skills – and in a competitive environment those skills come at a premium price. Third, it is clear that there will be greater opportunities for many salaried employees within the industry to reap the benefits of performance pay in a period of high activity.

Direct improvements to salary levels may also offset less positive changes to other conditions of employment. Company cars, sick pay schemes, access to medical benefits and the like are a corporate reference point that helps to define the position and 'insider' status of those who receive them. These benefits provide a badge of status and a measure of continuity where bonus earnings inevitably fluctuate. The availability of a company pension scheme is one of the factors that can shape long-term loyalty to an organisation, encouraging a feeling of being 'looked after' and enhancing retention in periods of prosperity. Final salary company pension schemes are nowadays much less common than they were and there has been a general trend across the private sector for them to be wound up, being retained only for existing, and typically more senior managers and the construction industry is not exempt from these changes (Hays, 2010). New entrants tend to be offered money purchase or defined contribution schemes and these changes have affected the construction sector. Barrett and Costain, for example, both closed their final salary schemes in 2009 (Wound up final salary schemes, 2010).

The relative gains for construction managers and professionals through until around 2008 undoubtedly reflect the wider business environment, where business opportunities put a premium on the skills of these occupations. This was also reflected in the working hours of these groups, with construction managers, in particular, working two-and-a-half or three hours a week more than their counterparts elsewhere in the economy. As a general principle, additional working hours for management and professional groups on site attract no additional payment. In this respect of course they are like salaried and professional staff in other sectors. It might be argued though that the extra working time does seem to have been factored in to overall improvements in their position over this period. This continued to be the case for all of the years in question, although the edge of uncertainty seemed to have been reached in 2009 with pay for some occupations – notably architects – beginning to show the pressures of the recession.

Directly-employed operatives

How do the arrangements and pay levels for operatives compare with other sectors? Given the level of increase in pay that has been noted for construction managers between 2004 and 2009, is there a comparable increase for operatives' earnings?

As stated above, the earnings of operatives are composed of hourly or weekly pay rates, sometimes reflecting the influence (the annual percentage pay rise), if not the actual rates, set out in multi-employer collective agreements. Different trades were clustered over time in a number of different national collective agreements within the construction industry – each one comprising at least one association representing employers' interests and a corresponding trade union or unions. By 2007 there were half-a-dozen major collective agreements in the construction sector in England and Wales, of which the largest, the Construction Industry Working Rule Agreement, was claimed (rather doubtfully) to affect working conditions for around 600,0000 operatives. A separate agreement for Building and Allied Trades, bringing together the Federation of Master Builders and the trade union, Unite, was said to cover 100,000 employees (Incomes Data Service, 2010: 24). Agreements for electrical contracting and plumbing (England and Wales) had a smaller constituency (50,000 and 30,000 respectively) whilst separate and smaller agreements persisted in some trades for Scotland and Northern Ireland.

As noted above, the impact of collective bargaining is less significant today than it has been in the past and claims that collective agreements 'affect' working conditions are at best tenuous. At the time of the last, authoritative survey of workplace employment relations (Kersley *et al.*, 2006: 119) only 13 per cent of private sector workplaces in construction recognised trade unions. Although national collective bargaining for the construction industry was said in 2010 (by a union representative interviewed for the case study below) to have survived, it is clear that the procedural arrangements governing the Construction Industry Joint Council (CIJC) have been weakened by the dissolution of the Construction Confederation at the end of 2009 and the future of this agreement must be in question. In practice employers in most of the construction industry (with the most notable exceptions being in electrical contracting and plumbing) determine wage levels without compliance with collective bargaining.

Earnings in the skilled trades within the construction sector generally kept pace with other comparable trades during the first decade of the twenty-first century. In **Table 10.2** below we provide data from the Annual Survey of Hours and Earnings for skilled trades and occupations and for construction and building trades on gross weekly pay in 2004 and 2009.

Table 10.2: Gross weekly pay of construction operatives 2004–09

Description	2004		2009		% Change 2004-09	
	£	£	£	£	£	£
	GWP* Median	GWP Mean	GWP Median	GWP Mean	GWP Median	GWP Mean
Skilled trades occupations	382.5	403.2	434.8	458.0	13.7	13.6
Electrical trades	453	477.4	516.5	548.7	14.0	14.9
Electricians, electrical fitters	463.2	484	551.1	571.2	19.0	18.0
Skilled construction and building trades	386.9	410.1	444.6	470.2	14.9	14.7
Construction trades	396.2	416.8	450.3	474.8	13.7	13.9
Steel erectors	428	453.4	439.4	457.5	2.7	0.9
Bricklayers, masons	379.9	404	434.4	455.9	14.3	12.8
Roofers, roof tilers and slaters	361	403.9	425.1	433.0	17.8	7.2
Plumbers, heating and ventilating engineers	450.4	463.2	512.4	530.3	13.8	14.5
Carpenters and joiners	376.5	400.2	446.3	473.4	18.5	18.3
Glaziers, window fabricators and fitters	326.2	343.1	372.5	401.0	14.2	16.9

Table 10.2: continued

Description	2004		2009		% Change 2004-09	
	£	£	£	£	£	£
	GWP* Median	GWP Mean	GWP Median	GWP Mean	GWP Median	GWP Mean
Construction trades	383.9	416.3	424.6	452.4	10.6	8.7
Building trades	349.2	380.5	410.4	445.7	17.5	17.1
Plasterers	352.5	411.9	409.5	443.3	16.2	7.6
Floorers and wall tilers	356.1	385.3	451.8	464.4	26.9	20.5
Painters and decorators	338.1	372.1	409.7	441.2	21.2	18.6
Operatives	367	398.9	419.3	452.3	14.3	13.4
Crane drivers	450.8	477.7	516.2	553.1	14.5	15.8
Fork-lift truck drivers	334.4	347.4	359.9	384.3	7.6	10.6
Labourers in other construction trades	309.5	303.7	341.5	360.1	10.3	18.6

*Data based on Office for National Statistics Annual Survey of Hours and Earnings (ASHE), 2004 and 2009b. * GWP = Gross Weekly Pay.*

Table 10.2 shows the pay position for the construction and building trades, relative to other skilled trades and occupations. The first point to note is the evidence of a hierarchy within the trades, with electricians, electrical fitters and plumbers (trades where collective bargaining continues to have some impact) better positioned than others. In 2004, both the median and the mean gross weekly pay for electrical and construction trades, including plumbers and also crane drivers, was higher than the figure for skilled trades occupations. The skilled construction and building trades saw their position improve overall between 2004 and 2009, broadly in line with the improvement for all skilled trades occupations, although their position peaked in 2008. By 2009, there was evidence of pay cuts in some trades, possibly because of reduced working hours. However the skilled trades experienced nothing like the increase in gross pay between 2004 and 2009 that benefitted construction managers and professionals. Significantly, the gross weekly pay for the skilled construction and building trades diminished when expressed as a percentage of the pay of construction managers from 64 and 66 per cent (median and mean respectively) in 2004 to 58 and 54 per cent in 2009.

Bonus or incentive payments constitute a further contrast between the management grades and those who are in receipt of hourly or job-based pay arrangements in some cases. Whilst an annual bonus or performance incentive is a significant and integral feature of pay structures in some firms for management grades, a weekly performance bonus may still form a part of the composition of earnings of hourly paid operatives. Unfortunately the ASHE data do not enable us to develop any further understanding of the contrasts that apply here and so we are unable to ascertain how many employees receive incentive pay.

Additional working hours may be required to ensure timely project progress and provision for an overtime calculator was a long-standing feature of national collective agreements for operatives. Hence overtime hours may provide additional financial compensation for additional hours worked. It is difficult to know how common this now is in practice as it seems likely that the application of a higher hourly calculator for the overtime rate is diminishing as firms look for and expect flexibility. In the manufacturing sector, there has been a trend toward 'single status' working and the harmonisation of terms and conditions of employment between salaried and waged workers. This has been encouraged by technological innovation, lean production management and associated changes in skills requirements in the latter years of the twentieth century (Druker, 2009: 102-3). However there is little evidence of comparable changes within the construction industry, where there is a continued differentiation between the position of management grades and craft and other skilled operatives.

The self-employed

Corporate provision sets the context for pay determination for managers and salaried staff and collective negotiation frames the historic pay systems (although no longer the pay) for craft and manual workers. However the self-employed are subject to unilateral pay determination. It is impossible to consider the question of 'reward' within the construction sector without reflecting on their position, since they are outside of the modest security and benefits of the directly employed. As stated above they may be paid by the day, by the task or by a combination of time and task. In most circumstances they lack job security, are reliant on networks for continuity of engagement and take responsibility themselves for insuring against sickness, providing for holidays and for provision of any other long-term benefits that they may need. Reflections on informality in construction (Chan and Raisanen, 2009) are pertinent here since the arrangements governing the engagement of self-employed workers and arrangements for their pay are difficult to document with any accuracy.

The self-employed phenomenon is long-established but has shown the capacity to adapt and mutate according to the framework of labour law and taxation regulations over at least the last 50 years. The growth in self-employment was documented above and many 'self-employed' workers are de facto employees. False self-employment arises where an 'engager' – in fact an employer – hires a worker as an independent 'sub-contractor', typically this will be through the Construction Industry Scheme. In this way he avoids employer's National Insurance Contributions which are due for an employee. The worker in this case pays National Insurance at a lower rate than other employees and benefits from preferential tax arrangements. He (for typically this is a male phenomenon) is unlikely to meet the criteria of true self-employment which would involve expenditure on and claims for tax deduction against provision of materials, plant or equipment. In reality of course he is unlikely to be incurring costs against such items. The construction industry has far more self-employed workers than any other industry and, despite initiatives since 1997, HM Revenue and Customs acknowledged in 2009 that some 300,000 workers were caught in the trap of bogus self-employment (Her Majesty's Revenue and Customs (HMRC), 2009: 12).

As a means of avoiding too close an association and thereby acquiring an obligation to observe employment regulations and a liability to payment of National Insurance, some companies arrange payment to the self-employed through specialist advisory firms, who function as intermediaries. These organisations sell their services to contractors or to sub-contractors by providing payroll facilities through which payment to the self-employed worker can be routed. In some cases large contracting firms are known to have set up their own 'agencies' to ensure that the relationship with workers is at arms' length. Accountants, lawyers and tax specialists develop a professional interest in defending the status quo and, as one union representative (interviewed whilst researching the case below) commented 'There is a multi-million pound industry in ensuring that workers don't have employee status'.

In all of these relationships, the individual worker is liable to find his own work – there is no real agency representation. He has little in the way of security or support in his engagement within the industry and indeed may find that this bogus self-employment is the only opportunity available to him (HMRC, 2009). It is not possible to document the terms and conditions under which he works, since his position is 'below the radar' of official statistics. Whilst anecdotal evidence suggests that illegal immigrants may be amongst those engaged in this way – and they may be particularly exposed to exploitative conditions – we have no hard evidence to this effect. What is clear is that the continued deployment of bogus self-employment is as much a part of 'reward' management within the construction sector as the practices associated with direct employment within the best known construction firms. A consultation on these arrangements was launched by the last Labour Government in July 2009 – an issue that remained to be addressed after 12 years of Labour government.

Case Study 10.1

Reward practices: MainCon and their sub-contractors

This case study provides an illustration of management practices on a high quality site in the south of England, in an area where the main contractor has a good reputation. It illustrates the range of reward practice to be found within one project. The case study was compiled from a small number of interviews – with the main contractor's construction manager, the project manager and the HR manager. Two sub-contractors were interviewed, representing very different perspectives in terms of the approach to employment and to 'reward', both of them contrasting their own approach with that of the industry more generally. There were no union representatives on site so a national representative from the Union of Construction Allied Trades and Technicians (UCATT) was also interviewed. Given the limited number of interviews, the picture presented here is necessarily partial and further work would undoubtedly be both interesting and challenging, since the world of sub-contracting – and sub-sub contracting – does not open up easily to researchers when the subject of employment contracts and payment arrangements is under discussion.

It appears that there is no one 'model' to offer for the ways in which reward is structured. It is difficult to claim that this project is 'typical' as company strategies vary, each contract is different, and sub-contractors provide a diverse array of organisational practices, in some cases focussed as much on tax avoidance as on 'reward' (or on the encouragement of tax avoidance as a form of enhanced reward). More extensive research would be necessary to probe the diverse and often highly informal reward practices. What is clear is that the discussion of reward – of payment systems and associated benefits – is a sensitive issue and one that requires more attention than it has received to date, from employers, from government and from researchers.

Reward practice within one construction project

'Paying a bad man bad money is not as competitive as paying a good man good money' (Sub-contractor)

The £35 million education project that forms the basis of this case study commenced in 2008 and was completed on time and to the client's satisfaction, 18 months later. At its peak, there were 248 people on site, of whom 85 per cent were sub-contractors. The main contractor, referred to here as MainCon, is a well-known and long-established company with a reputation for high quality work and a strong national and regional reputation. MainCon is unusual by the standards of the industry in having 'Investors in People' (IiP) status.

Maincon's project team was established at an early stage. The company nominated a highly experienced construction manager. He was supported by a project manager, two site managers, a surveying manager and two site surveyors + a trainee. In addition there was a site engineer + a trainee, a cladding manager, a mechanical and electrical consultant, a design manager and a site secretary.

Interestingly, both the Construction Manager and the Project Manager had begun their working lives with MainCon. The Construction Manager had been an apprentice carpenter 29 years earlier and had remained with the company throughout his working life. (On completion of this project, he was promoted to the position of Project Director). The Project Manager too had begun his working life with the company, serving a three-year period as an apprentice bricklayer, before promotion to charge hand and then trade foreman. Following successful study with the CIOB, he progressed further, though his service with the company was broken by spells elsewhere. Of course, the opportunities for personal progression that are suggested by this experience may not present themselves in quite the same way today, since apprentices are fewer in number than they were 20 or 30 years ago and the majority of management entrants are nowadays more likely to come through the graduate route. However MainCon has a strategic approach to 'growing its own' managers and there is currently a company-wide talent and succession process running through the organisation.

MainCon's managers had developed significant links within the region, having strong client bonds, built up over many years not only with the largest private sector company within the area but also with many of the local public sector bodies. Interestingly, the construction manager said that his role was concerned with the 'management of relationships' so that clients have confidence in the company and this was born out by the range of contracts won by the company within the area within a fairly short period. The interface with sub-contractors is acknowledged as important too, with managers indicating the significance they attach to ongoing relationships – not least because of the commitment which they ask for to their approach to site management and the management of health and safety on site.

Three distinct strata were evident within the framework of employment and reward practices on site but within each a wide array of practices was to be found.

The first group comprised salaried managers and professionals. Those employed directly by MainCon were paid on the basis of annual salaries with scope for an addition to pay through an annual performance bonus. The company has a 'people-focussed' CEO and this, it was felt, had led to significant recent progress in terms of HR practice. In many ways MainCon practices reflect 'new pay' theories, with performance-based pay rewarding both organisational and personal targets. Decisions on pay and benefits are taken by the Board's Remuneration committee for the top 15 – 20 employees but for other managers and professionals decisions emanate from the Executive Committee. They refer to an external construction salary survey and, in reflecting on that and on labour market information, will seek to retain a middling position in terms of salary levels. Salaries are situated within six broad bands as set out in **Figure 10.1** below:

Band 6 Entry point - Graduate Trainee
Band 5 - 6 Assistant Surveyor or Section Manager
Band 4 Site surveyor or Site Manager
Band 3ii) Project surveyor or Construction Manager
Band 3 i) Commercial lead or Construction Director
Band 2 Business Unit Director
Band 1 Managing Director .

Figure 10.1 Surveying and Construction salary bands

Performance-related bonuses comprise three strands: the largest being concerned with the company's performance targets and other strands concerned with personal targets and with health and safety. Targets are set by the Executive at the beginning of the year and personal objectives are reviewed at the year-end. Bonus payments may constitute one-third of annual salary and so represent a sizeable addition to basic pay.

Employees are encouraged to think in terms of 'total reward'. This means that they should understand something of the cost and the value of all of the benefits that they receive and there is some element of choice in selection of benefits according to personal circumstances. Each individual is issued with a 'total reward' statement. This comprises a car allowance scheme, life insurance, private health insurance, employer contribution to a money purchase pension scheme and flexible benefits to the value of £250 pa which can be used for, for example, dental cover. There are enhanced policies around maternity and paternity leave and the benefits, taken overall, constitute a significant addition to the value of the reward package.

Managers and professionals working for sub-contractors on site were rewarded on a salaried basis too – that is, they received annual salaries, pensions, holiday pay and car loans. This might operate through a formal structure – e.g. one sub-contractor pointed to hierarchy comprising senior surveyors, estimators, junior surveyors and contracts managers. Alternatively the focus may be on comparison with external and labour market rates

'If there's money around, people get rewarded ... we've got our ears to the ground' (Sub-Contract Director). But market rates may fluctuate downward in response to diminishing opportunity and tighter contract margins. The commitment to 'total reward' evidenced by MainCon in this case is distinctive within the industry and was not likely to be followed by sub-contractors.

Directly-employed construction operatives constitute the second group, employed either by MainCon or by sub-contractors on site. One sub-contractor reported a commitment to direct employment and application of the principles of the working rule agreement for the construction industry. This lays out hourly-based rates of pay, hours of work, holiday and sickness entitlement and additions to pay, for example for overtime or in the form of allowances. Operatives are eligible for bonus, for a stakeholder pension scheme and for enhancements to the hourly rate for additional skill – for example for banksmen. However the sub-contractor in question followed the agreement only in the most rudimentary way. The hourly rate paid exceeded the rates specified in the working rule agreement but some variation might apply over time – depending on the state of the economy and the opportunities for work in the locality – and that pay could go down as well as up. There was a flat hourly rate of pay with no provision for overtime or bonus. Operatives were expected to be available for a nine-hour day, which included travel time. So in general the national agreement was relevant in specifying minimum pay rates but for little else.

Directly-employed operatives are few in number and this is not the preferred route for most sub-contractors in engaging labour, although they may pay lip-service to employment through the working rule agreement. Nor do sub-contractors automatically reference the working rule agreement where they are employing direct. One sub-contractor termed this the 'old system', prone in the past, he said, to strikes and union disruption. An alternative approach to pay determination was geared to ensuring that individual operatives were motivated and hardworking, with the maxim that 'you pay the extra money if he produces'. 'There will be a base rate, but then if there's someone who's really good you'd pay him over the rate' (Sub-contractor). The arrangement seems to reference a pay-range within which particular trades were likely to be paid although ultimately it was a form of employment that was almost entirely individualised.

MainCon's Project Manager felt that the directly employed represented value for money for the main contractor, contrasting their hourly rate of pay with the £14 per hour that might be required for agency labour. Agency staff were used for holiday cover or for specialist work – but Maincon did not employ many operatives direct and those who were employed were not, on the whole, from a trades background.

For the final group, self-employed operatives engaged by sub-contractors, conditions were less formally regulated. Described as a 'risk/control' system, the conditions of the sub-contract workforce are often trade specific and dependent on the tasks required when work is sub-let. A wide spectrum of payment arrangements might apply but in general the self-employed engaged by sub-contractors were paid by the task rather than by the hour, encouraging early task completion. In some cases there may be a combination of time and task requirements. Work might be priced by the sub-contractor and sub-let so that the people who are taken on are paid a day rate for completing that particular piece of work and sometimes they in turn are referred to as 'sub contractors'. Sub-contract trades included shuttering carpenters, steelfixers and concrete mix. The day-rate for the self-employed reflects the 'going rate' for each particular trade and sub-contractors are alert to the need to keep their payments aligned with others since an apparently modest increase in day-rates could lead to a significant overall rise in costs. Unlike directly-employed operatives, 'self-employed' workers had no job security and were engaged solely when required and retained only for so long as the job in question lasted. In consequence they would have no prospect of progression – other than through the relationships that might be forged on site and which might lead to the next opportunity of work. The foremen on site for the main sub-contractors had their own networks which were deployed when required – for example if the sub-contractor's directly-employed workforce was too distant or already engaged elsewhere. Main sub-contractors pay those working for them and the 'self-employed' are then expected to take personal responsibility for their own tax payments. There was no evidence that agencies were used and neither the sub-contractors nor the Project Manager saw advantages in doing so.

Discussion and conclusion

A complex and segmented picture of reward practices in the industry is painted by the evidence, both from published sources and from the Case Study. If a strategic approach to reward management is characterised by a link between organisational requirements and individual reward and by the development of more careful attention to each individual component within the reward package then there is some evidence within our high quality main contractor's treatment of its managerial and professional strata of a 'new pay' or strategic approach. Here we can see the investment in careful recruitment and training, development and progression opportunities and a structured approach to reward, with features that take account both of individual preferences and of organisational requirements. With support from its human resource management department and leadership from the Chief Executive, it seems that 'new pay' arrangements may be relevant to some large contracting organisations, although we cannot deduce that it is adopted by all or even many of them.

The disproportionate improvement to the pay of construction managers between 2004 and 2009 is an interesting finding to emerge from the ASHE data. It might be argued that others within the industry benefitted from the good years leading up to the recession but the earnings of construction managers increased beyond the level of skilled trades or even of professionals and ahead of earnings for other comparable occupations, suggesting that their importance to project delivery has been recognised by their employers. The disproportionate improvement to construction managers' pay cannot simply be ascribed to the level of economic activity since this might be expected to impact upon the earnings of those in the construction trades too. It may be that the role and responsibilities of construction managers in managing relationships is attracting greater recognition and a pay premium given the complexity of managing inter-organisational relations and addressing client pressures. It seems likely that this would be the case, particularly for those with more experience and standing. This would be an interesting development but of course account must be taken of counter-veiling factors – especially the extensive working hours without overtime payment that may be required of this group in a period of intensive activity and the possibility that some valuable benefits (especially pensions) may have been curtailed. It may also be the case that the pay of construction managers is more volatile than for other groups. In 2010 it was reported that 96 per cent of firms had frozen pay and for those changing jobs basic salary overtook 'finding engaging work' as the priority (Hays, 2010). A longer-term perspective is needed to be sure of trends but it would not be surprising if the more discerning employers were giving greater recognition to the importance of the management role.

The future of multi-employer collective bargaining is in question following the demise of the Construction Confederation. National bargaining has had such limited relevance in recent years that it is difficult to see how it can survive the disappearance of the largest employers' organisation unless there is some co-ordinated initiative to sustain it. This in turn raises questions about the choices that employers may make in the future. Will there be greater interest in single status working and harmonisation of terms and conditions for manual workers with those of their non-manual counterparts, as has happened in manufacturing? Or will employers have greater recourse to bogus self-employment?

In reviewing the ASHE statistics and literature about the industry and considering too the ways in which organisational boundaries are re-drawn (Grimshaw *et al.*, 2005b) and the evidence emerging from our (admittedly limited) case study it seems that in considering reward management we cannot avoid practices associated with sub-contracting and bogus self-employment. 'Reward' must be considered in the context of the complex framework of inter-organisational relations. In the multi-organisational workplace that is the typical construction project, sub-contracting and informal self-employment arrangements offer an alternative to the formal-ised terms and conditions that accompany the conventional employment contract. Bogus self-employment and systematic tax avoidance – and in some cases tax evasion – enable the payments made to sub-contractors to be augmented by tax advantages. Perhaps as significantly, they relieve the 'employer' from the obligations of meeting the employer's National Insurance costs and employment responsibilities. The workforce engaged in this way may be deprived of security and of the long-term benefits of employment (with the consequent long-term cost to the Exchequer). We have no evidence to offer about the motivational implications or the 'psychological contract' of these informal working arrangements. Whether this is, currently and in the longer-term, to the benefit of the higher quality contractors and their clients – or indeed to the public purse – is a matter that deserves further attention.

It is to be expected that the construction industry would have distinctive reward practices but the scale of the false self-employment that is reported is remarkably high, both by comparison with other UK industries and by contrast with international comparators. To date there is little evidence that it benefits the industry by enhancing motivation or productivity. It seems odd that successive governments have failed to grapple with this issue, given the potential benefits for the public purse if they were to do so successfully. It seems odd too that there are contractors and sub-contractors who are intent on working to the highest standards who have not taken more active steps to regularise employment and taxation arrangements on the projects which they control. Although it is never practicable to transplant employment policies and regulations from one country to another, influences from mainland Europe where there are more formal and regulated structures suggest that there are values and influences which could be given further consideration.

Taken together, the factors above highlight the unusual and distinctive shape of reward practice in the context of the complex employment and contractual structures that characterise the construction industry. The most developed reward strategies are to be found at the level of the firm (at least of particular firms) but the industry as a whole could benefit from further research to evaluate and review the costs and the benefits of current reward practices within the sub-contracting process. Many firms could benefit from a thorough-going examination of the approach that they adopt in order to meet the challenges of the post-recessionary period and to avoid over-reliance on inherently unstable forms of labour engagement. These issues interlock with questions of skill shortage and training opportunity and there is scope for a more long-sighted and holistic approach to the management of the employment relationship and the rewards that it brings for all parties.

References

Armstrong, M. and Murlis, H. (2007) *Reward Management: a handbook of remuneration strategy and practice*, revised 5th edn, Kogan Page, London and Philadelphia.

Brown, Wilfred (1962) Piecework Abandoned: the effect of wage incentive systems on managerial authority, Heinemann, London.

Chan, P. and Raisanen, C. (2009) 'Editorial: informality and emergence in construction', Construction Management and Economics 27: October. 907–12.

Construction Industry Joint Council (2006) Working Rule Agreement for the Construction Industry, Construction Industry Joint Council, London, 2006.

Dainty, A., Bagilhole, B. and Neale, R. (2001) Male and female perspectives on equality measures for the UK construction sector, Women in Management Review 16(6), 297–304.

Dainty, A., Raidén, A. and Neale, R. (2004) Psychological contract expectations of construction project managers, Engineering, Construction and Architectural Management 11(1), 33–44.

Druker, J. (2007) Industrial relations and the management of risk in the construction industry, Dainty, A., Green, S. and Bagilhole, B. (2007) (eds), People and Culture in Construction, Taylor & Francis, London.

Druker, J. (2009) Determining pay, White, G. and Druker, J. (eds) (2009) Reward Management: a critical text, Routledge Studies in Employment Relations, Routledge, London and New York, 100–119.

Gospel, H. and Druker, J. (1998) The survival of national collective bargaining in the electrical contracting industry: a deviant case?, British Journal of Industrial Relations, 36 (2) June, 249–67.

Grimshaw, D., Marchington, M., Rubery, J. and Willmott, H. (2005a) Introduction: Fragmenting work across organisational boundaries, Marchington, M., Grimshaw, D., Rubery, J. and Willmott, H., Fragmenting Work: blurring organisational boundaries and disordering hierarchies, Oxford University Press, Oxford, 1–38.

Grimshaw, D., Marchington, M., Rubery, J. and Willmott, H. (2005b) Redrawing boundaries, reflecting on practice and policy, Fragmenting Work: blurring organisational boundaries and disordering hierarchies, Oxford University Press, Oxford, 267–87.

Guest, D. and Conway, N. (2001) Public and Private Sector Perspectives on the Psychological Contract: results of the 2001 CIPD Survey, Chartered Institute of Personnel and Development, London.

Hays 2010 Construction Salary Survey, 2010 www.building4jobs.com/careers/salary-surveys.316512.hays-2010-construction-salary-survey (accessed 24 March 2011).

HM Revenue and Customs (2009) False Self-employment in Construction: taxation of workers, HM Treasury, London.

Hutton, W. (2011) Hutton Review of Fair Pay in the Public Sector; final report, 2011, Crown Copyright, London.

Incomes Data Services (2010) Pay Report 1057, September.

Kersley, B., Alpin, C., Forth, J. Bryson, A., Bewley, H., Dix, G. and Oxenbridge, S. (2006) Inside the Workplace: findings from the 2004 workplace employment relations survey, Routledge, London and New York.

Korczynski, M. (1997) Centralisation of collective bargaining in a decade of decentralisation: the case of the engineering construction industry, Industrial Relations Journal 28(1), 14–26.

Lawler, Edward, Third, the (2000) Rewarding Excellence: pay strategies for the new economy, Jossey Bass, San Francisco.

Leat, M. (2003) The European Union, Hollinshead, G., Nicholls, P. and Tailby, S, (eds) Employee Relations, Prentice Hall, London and New York.

McGrath-Champ, S. and Rosewaarne, S. (2009) Organisational change in Australian building and construction: rethinking a unilinear 'leaning' discourse, Construction Management and Economics, November 27: 1111–28.

Mayo, E. (1933) The human problems of an industrial civilisation. Macmillan, New York. Cited in Steers, R. M., Porter, L. W. and Bigley, G. (1996), Motivation and Leadership at Work, McGraw Hill, New York, 28.

Mishel, L., Bernstein, J. and Shierholz, H. (2009) The State of Working America, 2008–09, Economic Policy Institute, ILR Press (an imprint of Cornell University Press), Ithaca and London.

ONS (Office for National Statistics; 2004) Annual Survey of Hours and Earnings: analysis by occupation www.statistics.gov.uk/downloads/theme_labour/ASHE_2004_inc/2004inc_occ4.pdf (accessed 4 August 2010).

ONS (2009a) Construction Statistics Annual, 2009, Table 12. www.statistics.gov.uk/downloads/theme_commerce/CSA-2--0/Operning-page.pdf (accessed 16 August 2010).

ONS (2009b) Annual Survey of Hours and Earnings: analysis by occupation (revised), Table 14a, www.statistics.gov.uk (accessed 23 March 2011).

ONS (2010) Labour Force Survey Historical Quarterly Supplement Table 10: Employees and Self-Employed by Industry Sector, ONS, London.

Pensions Act 2008 (2008) Crown Copyright. www.legislation.gov.uk/ukpga/2008/30/pdfs/ukpga_20080030_en.pdf (accessed 24 March 2011).

Perkins, S. and White, G. (2008) Employee Reward, Alternatives, Consequences and Contexts, London, CIPD.

Roethlisberger, F. and Dickson, W. J. (1939) Management and the worker, Harvard University Press, Cambridge, Mass. Cited in Steers, R. M., Porter, L. W. and Bigley, G. (1996) Motivation and Leadership at Work, McGraw Hill, New York, 28.

Rubery, J. and Earnshaw, J. (2005) Employment policy and practice: crossing borders and disordering hierarchies, Marchington et al. (2005) Fragmenting Work: blurring organisational boundaries and disordering hierarchies, Oxford University Press, Oxford, 157–77.

Sang, K., Ison, S. And Dainty, A. (2009) The job satisfaction of UK architects and relationships with work–life balance and turnover intentions, Engineering Construction and Architectural Management 16 (3), 288–300.

Schuster, J. R. and Zingheim, P. K. (1992) The New Pay: linking employee and organizational performance, Lexington, New York.

Sparrow, P. (2009) International reward management, White, G. and Druker, J. (eds), Reward Management: a critical text, Routledge Studies in Employment Relations, Routledge, London and New York, 233–57.

Standing, G. (1997) Globalization, labour flexibility and insecurity: the era of market regulation, European Journal of Industrial Relations 3(1), 7–37.

Storey, J (2007) Human Resource Management: a critical text, 3rd edn, Thomson, London.

Taylor, F. W. (1913) The Principles of Scientific Management, New York and London, Harper & Brothers.

Taylor, S. (2009) Occupational pensions, White, G. and Druker, J. (eds; 2009) Reward Management: a critical text, Routledge Studies in Employment Relations, Routledge, London and New York, 192–211.

White, G. (2009) Determining pay, White, G. and Druker, J. (eds; 2009) Reward Management: a critical text, Routledge Studies in Employment Relations, Routledge, London and New York, 23–48.

Wound-up final salary schemes (2010) www.cookham.com/community/equitable/finalsalaryz.htm (accessed 23 March 2011).

Index

294   *Index*